Snowdonia

M. F. HOWELLS,
B. E. LEVERIDGE
and A. J. REEDMAN

ROCKS & FOSSILS
A geological field guide

London
UNWIN PAPERBACKS
Boston   Sydney

First published in Unwin Paperbacks 1981

UNWIN® PAPERBACKS
40 Museum Street, London WC1A 1LU

© M. F. Howells, B. E. Leveridge and A. J. Reedman 1981

British Library Cataloguing in Publication Data

Howells, Malcolm Fletcher
  Snowdonia. − (Rocks and fossils).
  1. Geology − Wales − Snowdonia
  I. Title   II. Leveridge, B E   III. Reedman, A J
  IV. Series
  554.29′25   QE262.S/

ISBN 0-04-554005-5

Set in 9 on 11 point Times by Typesetters (Birmingham) Ltd., and printed in Great Britain by Hazell Watson & Viney Ltd., Aylesbury, Bucks.

Invited foreword

Over the past few years there has been increasing public interest in the geological sciences. This derives from an increased awareness of their twofold role: first, they sketch out the history of our planet, and secondly they help to provide the mineral resources on which modern society depends. Although this spread in interest is to be welcomed, it can lead to certain undesirable consequences as more and more visitors come to examine a finite number of instructive rock exposures.

The authors and publishers of this book are fully aware of the need to make the best possible use of the outcrops it describes and have accordingly consulted with the Nature Conservancy Council, the official body responsible for conservation in Britain. All their efforts however will be nullified if today's readers choose to ignore their responsibility to students of the future.

It must be emphasised that the vast majority of geological outcrops are in private hands and that access to them is through the goodwill of the owners and occupiers. If lost, this goodwill will be difficult, if not impossible, to regain — it can take only one careless act to cause offence. Further, many geological localities can lose their interest through the cumulative effects caused by the unnecessary use of hammers. To observe and record, collecting any necessary specimens only from fallen rock, will in general give a better understanding of geology than will a physical assault on selected portions of the rock face. The indiscriminate use of hammers is thus as profitless as it is damaging.

The authors and publishers of this guide have done their best to ensure the maximum benefit from your visit to Snowdonia; it remains for you to ensure that the same benefit remains available to your successors.

DR G. P. BLACK
Nature Conservancy Council

Preface

We have had the good fortune to work in the hills and mountains of Snowdonia for several years making detailed geological maps. Our contacts with many of the increasing numbers of hill walkers and student groups in the area have indicated a wide and enthusiastic interest in the rocks that make up the hills, and for this reason we were stimulated to write this book.

The processes that operated during the geological evolution of the Snowdonia area were many and varied, and the time span over which they operated was vast. Many people are aware that the ice age played a role in shaping the area but are surprised when it is explained that this event was geologically very recent and, when considered with the catastrophic events that preceded it, almost insignificant. There is no better way of appreciating this than seeing, in the course of a few hours' walking in the hills, marine fossils in rocks now forming some of the highest ground in the area and products of violent volcanic eruptions that took place 500 million years before the Ice Age.

The book briefly describes the geological background of the area and presents excursion itineraries to examine many of the interesting features of its geology. For the beginner, there are introductory sections defining terms and describing geological processes, some knowledge of which is necessary for an appreciation of the geological story revealed by examination of the rocks. The outcrops described in the excursions will also be of interest to advanced students and professional geologists wishing to exercise their own interpretative skills.

The first geological survey of Snowdonia, on the 1-inch scale, was published in 1852. Our work in the area has been on a second, more detailed survey, and in writing this book we have drawn on experience gained during this survey. We are grateful to the Director of the Institute of Geological Sciences for permitting us to collaborate in writing this book and to use official photographs and some sketch maps based on our work.

There is an immense wealth of interesting geological localities in the area, and we should like to think that some readers will be sufficiently stimulated to explore in more detail, using the published geological maps.

M. F. HOWELLS, B. E. LEVERIDGE, A. J. REEDMAN
Institute of Geological Sciences, Leeds

Contents

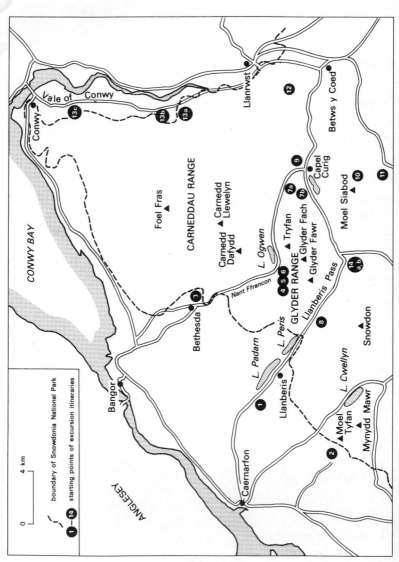

Frontispiece General location map.

Part 1: Geological background

1 Introduction

The sharp contrasts and splendour of the scenery of North Wales are closely related to the geological process. Nowhere in this area is the landscape more impressive than in Snowdonia, most clearly displayed in the profile that it presents to visitors approaching from the east across the Denbigh Moors or from the much denuded landscape of Anglesey. In this severely dissected topography the rocks are very apparent and, as a result, have attracted generations of geologists and students. This field guide is intended for the interested layman and student, to provide geological information along a number of tracks and walks in northern Snowdonia, together with discussion of some of the processes that formed the rocks and shaped the landscape. The excursions described are within the area between the Conwy Valley and Llanberis to the west and from the Carneddau south to Snowdon.

Wales holds an important place with regard to the development of geology as a science. In particular, it was here that the system boundaries of the Lower Palaeozoic were defined by Lapworth after a prolonged controversy between Sedgwick and Murchison in the nineteenth century. At this time the Geological Survey produced the first geological maps of the area on the scale of 1 inch to the mile. Towards the end of the century detailed examination of the microscopic character of the rocks of Snowdonia began with the work of Harker, but it was not until the field studies of Williams in the 1920s that a comprehensive account of the stratigraphy and petrography of the rocks of the Snowdon area was published.

The rocks of Snowdonia are mainly of Cambrian and Ordovician age, partly sedimentary and partly volcanic in origin (Fig. 1). Following their deposition the rocks were affected by prolonged earth movements, which elevated them high above the present relief of Snowdonia. In this position they were subjected to continuous erosion, and through geological time the hills were denuded, submerged and uncovered, so that by the time of the Ice Age (Pleistocene) they were approximately at their present scale. During the Pleistocene glaciation the hills were moulded by the ice into their present form.

It is hoped that the excursions will, in particular, stimulate interest in the manner in which a geologist is able to interpret volcanic rocks in ancient sequences and to relate these to the sedimentary rocks with which they are interbedded. In addition, attention will be drawn to the erosional and depositional effects of the glaciation, which are so striking throughout the area.

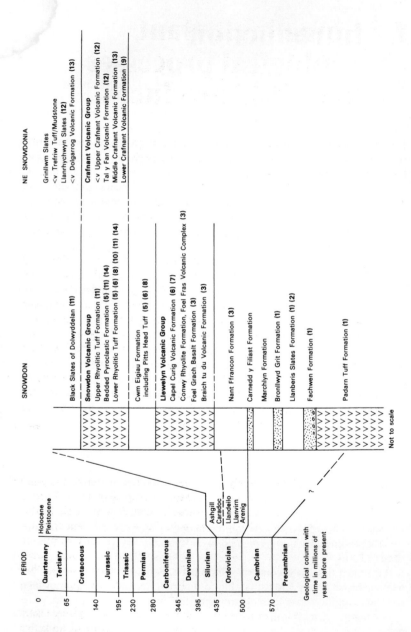

Figure 1 Detailed succession in northern Snowdonia, related to the geological column. Numbers relate to excursions.

2 Some important geological processes and their products

In order to interpret sequences of ancient rocks satisfactorily there are two basic geological tenets that are worth consideration. The first was developed in the seventeenth century and is referred to as the **law of superposition**. This states that in a sequence of sedimentary or extrusive volcanic rocks the youngest bed is at the top and the oldest at the base. However, to anyone who has crossed a mountain chain such as the Alps it is immediately apparent that the law only stands when the beds have not been severely folded.

The second tenet is that of **uniformitarianism**, first clearly defined, by James Hutton, in the late eighteenth century. It states that geological processes that operate at present also operated in the past and will continue to do so in the future. The principle is ingrained in geological thinking, and it is a standard practice for geologists interpreting ancient sequences to look for possible present-day analogues. For obvious reasons this is easier to do in some cases than in others; more is known of terrestrial and shallow water sediments than of deep oceanic sediments, for instance. Similarly, more is known of land-sited volcanism than of that occurring in the oceans. However, with this principle in mind we shall first consider, in general terms, some of the important geological processes and their results and relate them to the geological sequence in Snowdonia.

Volcanic rocks

Volcanic rocks originate largely from new molten material (magma) formed deep beneath the Earth's surface. Magma erupted on to the Earth's surface is called **lava**, and the rocks formed when magma cools in this situation are known as **extrusive igneous rocks**. However, magma does not always reach the surface of a volcano; it may cool and solidify within the earlier rock cover to form **intrusive igneous rocks**. Volcanic rocks composed entirely of fragments, either of magma or of older rock, that have been ejected from the volcano form another category called **pyroclastic rocks**.

Magma consists of a combination of molten silicates and gases. Gases escaping from magma are often the most conspicuous feature of an active volcanic area; by the time the magma has cooled and the silicates have

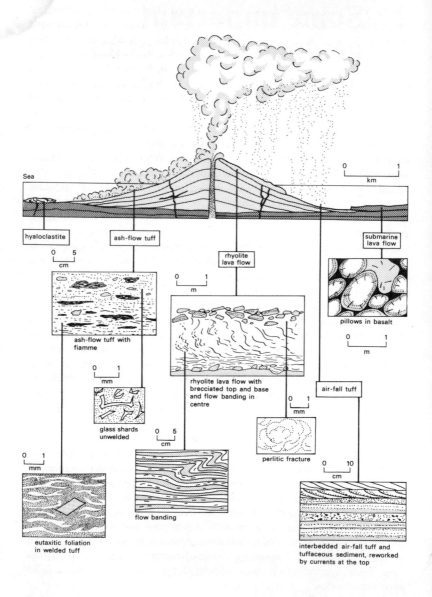

Figure 2   Diagrammatic representation of a volcano, showing some of the more common processes and related rock fabrics.

formed solid rock, most of the gas has escaped. The mineral and chemical composition of the rock can be determined by means of microscopic examination and chemical analysis. The simplest classification of these rocks is based on the amount of light and dark-coloured minerals. Rocks such as **rhyolite** and **granite**, which are rich in light-coloured minerals (feldspar and quartz), are referred to as **acid**. Rocks such as **basalt** and **dolerite**, which are rich in dark-coloured minerals containing iron and magnesium, are referred to as **basic**. Of course, there is a complete gradation of magma compositions, so that there is also a much-used category of 'in between' rocks termed **intermediate**.

Pyroclastic rocks reflect the wide range of materials that can be erupted as fragments during the escape of gas from a volcano (Fig. 2). The extremely fine fragments (<2 mm diameter) often included in the extensive clouds associated with volcanic activity are referred to as **ash**, which, when consolidated into a rock, forms **tuff**. Slightly larger fragments are called **lapilli** (2–64 mm) and **blocks** (>64 mm). Large lumps of magma ejected while still plastic are called **bombs**. Rocks composed of large fragments are termed **volcanic breccias** or **agglomerates**, depending on whether the materials that they contain are considered to have been ejected in a solid or a plastic state respectively. Generally, these pyroclastic rocks are the consolidated blanket of debris formed by the simple gravitational fall of ejected fragments on to the Earth's surface. Fine ash particles are commonly transported long distances by winds before settling to form **air-fall tuffs** (Fig. 2). The air-fall debris may fall directly on land or into the sea, where a submarine deposit is formed. When the pyroclastic debris (>25%) becomes admixed with epiclastic sediment (p. 16) (>25%), the resulting rock is termed **tuffite**.

In some instances, as the magma rises in the volcano, small gas bubbles are formed in the molten liquid as pressure is relieved, until at the point of eruption the rapidly expanding gas bubbles cause the magma to froth and disintegrate. The disrupted material may be erupted as a hot **ash-flow**, which consists of small fragments of the disintegrated magma suspended in gas. The fragments, derived from the walls of the gas bubbles, are distinctively curved and Y-shaped and are termed **shards** (Figs 2, 3d). They consist of **volcanic glass**, which results from magma cooling so rapidly that the silicates do not form crystals. Where the fragmentation is incomplete, fragments of frothy volcanic glass (**pumice**) occur. Crystals may also be present in the ejected material if the magma started to crystallise prior to the eruption. The ash-flows may travel for several kilometres from their point of eruption. Their deposits are called **ignimbrites** or **ash-flow tuffs** (Fig. 2). In places the fragments are hot enough when deposited from the flow for the shards to stick together and flatten and for the pumice fragments to collapse under the weight of

Figure 3 Photomicrographs of volcanic rocks (IGS photographs).
    (a) Accretionary lapillus. Core of shard fragments and vitric dust with fine-grained rim. Plane polarised light (×40).

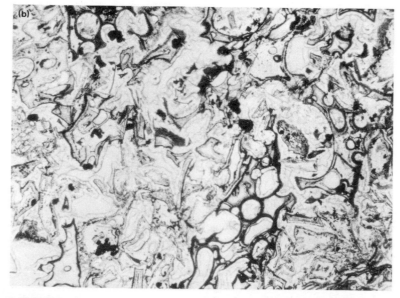

(b) Hyaloclastite. Fragments of vesiculated basaltic glass replaced by chlorite, some margins accentuated by iron oxide grains. Plane polarised light (×40).

(c) Rhyolite. Flow-orientated feldspar microlites with a few feldspar phenocrysts, in a devitrified glassy matrix. Crossed nicols (×45).

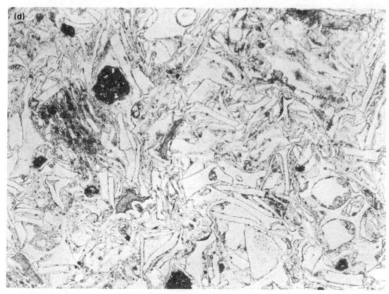

(d) Vitroclastic texture. Devitrified shards and pumice fragments in an acidic ash-flow tuff. Plane polarised light (×30).

(e) Eutaxitic texture. Devitrified shards and pumice deformed about crystals (C) and lithic fragments (R). Plane polarised light (×40).

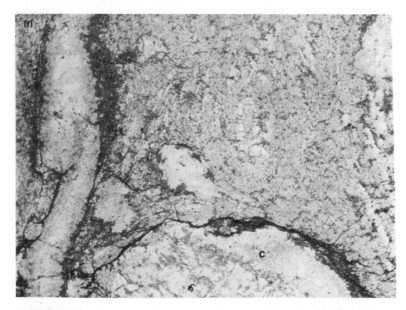

(f) Parataxitic texture. Extreme flattening of shards adjacent to an altered feldspar crystal (C). Plane polarised light (×60).

accumulated material; the resulting tuffs are referred to as **welded tuffs**. Since the original recognition of welded tuffs in New Zealand they have been described from many volcanic areas in rocks previously regarded as lavas, and as we shall see, North Wales is no exception.

Having examined some of the material produced by volcanic activity, it is necessary for us to describe briefly some of the structures that they are known to have formed in areas of recent activity, before we consider the North Wales area. To most people the term **volcano** directly relates to the large, symmetrical, often elegant, volcanic **cone** (Fig. 2) topped by a circular depression **(crater)** within which is found a **vent** − the opening through which the volcanic material issues to the Earth's surface. Of this type, the Mayon volcano in the Phillipines and Mount Fuji in Japan are among the most well known. However, many volcanoes, such as those of the Hawaiian Islands, have extremely low profiles resembling upturned saucers; these are referred to as **shield volcanoes.**

The form of structure that occurs about the vent depends largely on the types of material produced and the rate at which they are erupted. The steep-sided cones normally consist of alternating layers of solidified lava and pyroclastic debris, whereas the shield volcanoes consist mainly of basaltic lavas, which are extremely fluid and mobile when molten. Acid lavas are much stiffer and more viscous and do not flow so easily. The central parts of acid lava flows often contain narrow **flow bands** of contrasting texture and mineral composition, which reflect the original movement of the lava flow (Figs 2, 3c). The flow banding frequently develops parallel to the base in the lower part of the flow, but it may become contorted higher up and pass into a breccia zone at the top. This breccia represents the fragmentation of the solid skin forming at the surface of the lava due to continued flow movement within its still plastic centre (Fig. 2). Brecciation also occurs at the flow front, and blocks are incorporated into the base of the flow as it advances. Because of their high viscosity the acid lavas may also form steep-sided **volcanic domes** around the vent − a common feature of the Auvergne region in central France and the volcanoes of the West Indies. In places, great spines of viscous acid magma protrude above the surface of the domes. Such a phenomenon occurred in 1902 in the crater of Mont Pelée in Martinique, following a catastrophic eruption during which an ash-flow travelled down a valley from the summit and an associated hot gas cloud was projected, at great speed, out of the confines of the valley, to devastate the town of St Pierre and destroy all but two of its 30 000 inhabitants.

Volcanic activity is not always centred on the summit of a volcano. On a large volcano, such as Mount Etna in Sicily, numerous small cones occur across the whole structure, and some of the recent activity stems from a series of vents aligned along a fissure. At many times in the Earth's history extremely fluid basaltic lavas have been erupted on to the surface through a series of fissures to form vast basaltic plains, such as the Deccan plateau in India, with the complete absence of a central volcanic structure. The only volcanic rocks other than basalts that form extensive plains are ash-flow tuffs, such as those occurring in North Island, New Zealand.

Volcanoes do not occur exclusively on land. Many erupt beneath the sea, and a feature of volcanic activity that is of particular relevance to the study of volcanic rocks in North Wales is the effect of water on eruptions. Most of the sediments interbedded with the volcanic rocks in North Wales are regarded as having been deposited in a marine environment, and it is reasonable to expect that water at some time influenced both the nature of the contemporaneous volcanic eruptions and the rocks produced. It was recognised by some of the earliest volcanologists that many of the oceanic islands are but the tips of submarine volcanoes built up from the ocean floor. In 1963 the emergence of the volcanic island of Surtsey, off the southern coast of Iceland, provided a dramatic demonstration of this activity. In the early stages of the development of Surtsey, water was able to enter the vent directly, and the eruptions were of a particularly distinctive form, with huge clouds of ash and steam. The eruptions were sufficiently voluminous for the volcano to be built up above sea level, forming an island, even though the debris was continuously being eroded by the surrounding sea. Part of the activity in the later stages produced basaltic lava, which formed a solid shell around much of the unconsolidated ash and without which it is unlikely that the island would have survived. Much of the ash that fell into the sea was subjected to wave action, and the process of sedimentary sorting began immediately, with sediments composed of volcanic material (**volcanogenic sediments**) being produced. When a large volume of ash rapidly accumulates in the sea on the slopes around the vent, it may become waterlogged and unstable; this produces mass collapse of the deposits, which flow or slide into deeper water. As we shall see later, some geologists now believe that hot ash-flows erupted on land can flow into the sea, maintaining their identity, and deposit submarine ash-flow tuffs, which may subsequently weld.

Until recently the behaviour of lava flows under water had not been observed. However, with the rapid growth of the investigation of the ocean floor much more information is now forthcoming. Most commonly, in a marine environment the flows are of basalt, and characteristically they contain large ellipsoidal bodies called **pillows**, which are rimmed by a glassy skin and show radial joint patterns (Fig. 2). The development of these pillows has now been observed by an American volcanologist, Jim Moore, who filmed a basalt flow extending into the sea off the Kilauea volcano in Hawaii. Tongues of lava extended through the rubble at the front of the flow, their skins fractured, and the hot lava within the tongues protruded through the fractures like toothpaste from a tube. These protrusions became dislodged on the steep flow front and formed pillows, some elongate, others more rounded.

This process of pillow development at basalt flow fronts is commonly accompanied by the formation of a deposit of glassy fragments, termed **hyaloclastite** (Fig. 3b), by the flaking off of the chilled glassy skins of the pillows. Hyaloclastites of this type occur only in local accumulations, but others form extremely thick deposits not clearly related to basalt flows; it is likely that these develop by a process involving the explosive eruption of basaltic magma fragments into water. The surfaces of the hot basaltic fragments chill rapidly in contact with water, the entrapped gases expand, and the fragments break, exposing a fresh glowing surface and allowing a repetition of the process.

Ancient volcanic rocks, such as those found in North Wales, have invariably been altered to some degree as a result of: (a) their burial below younger rocks; (b) subsequent subjection to stress by earth movements, during which they were folded, faulted and cleaved; and (c) weathering. Of these, the first and second processes have caused the greatest mineralogical changes. The volcanic glass has **devitrified,** becoming finely crystalline; feldspars have changed composition and in places been completely replaced; the iron and magnesium minerals commonly have lost their original form and been replaced by fine mineral aggregates. Locally, textures have been modified by **cleavage** – very closely spaced parallel sets of planes along which the rock tends to split. Because of these alterations the rocks are not always easy to interpret. However, with careful investigation it is possible to unravel many of the original textures and compositions, the form of the rock units and the distribution of the eruptive centres. In fact, it should be pointed out that in the study of ancient volcanic centres it is possible to examine the deeper regions of the original structures, whereas modern volcanologists have to satisfy themselves with examination of just the skins of the structures – and even this in often hazardous conditions.

The earliest geological surveyors in North Wales recognised that many of the rocks in Snowdonia were volcanic. They differentiated them from the sedimentary rocks and descriptively called them 'greenstone', 'felstone' and 'porphyry', implying that they were mainly intrusive rocks, although Adam Sedgwick, writing in 1843, thought that most were extrusive. The first comprehensive mapping, on a scale of 1 inch to 1 mile, was completed in the 1850s by the Geological Survey, and the area was described by Ramsay (1881) in the accompanying memoir. In his account many of the acid volcanic rocks were described as rhyolite lavas emplaced under water.

Later, Harker (1889) examined the Ordovician volcanic rocks in Caernarvonshire and defined both rhyolite lavas and intrusions, basic intrusions and also a group of intermediate lavas termed **andesite**. Harker recognised that many of these rocks had suffered mineralogical changes following their emplacement and discussed the origin of the siliceous nodules that commonly occurred. He supported Ramsay's conclusion that the extrusive rocks were emplaced in water.

In 1905 Dakyns and Greenly recognised bedded submarine acid tuffs and, more importantly, massive unbedded acidic rocks that locally showed pyroclastic textures. Greenly was aware of the similarity of these rocks to the products of the glowing ash flows (**nuées ardentes**) erupted violently from the volcanoes of Soufrière on St Vincent and Mont Pelée on Martinique in 1902, although he concluded that the North Wales rocks were emplaced under water. As with many astute observations, Greenly's was not immediately pursued and in the following years, although small parts of the geology of North Wales were described in detail by various geologists, the volcanic rocks received scant attention. An exception was the outstanding work of Howel Williams (1927), who described the volcanic rocks of Capel Curig and Snowdon in considerable detail. He recognised a wide range of volcanic breccias and tuffs in the sequence and concurred with Greenly on the possible ash-flow origin of some of the massive tuffs. As a result of their association with marine sediments, he continued to regard the emplacement of the volcanics as subaqueous.

There was a major reappraisal in the 1950s, when Oliver (1954) and Rast, Beavon and Fitch (1958) determined that many of the rocks previously thought to be acid lavas were, in fact, welded ash-flow tuffs, or ignimbrites. However, on its original definition, ignimbrite was considered to form exclusively in a subaerial environment. Beavon, Fitch and Rast (1961) persuasively reconciled this concept with Snowdonian stratigraphy by postulating repeated emergence of volcanic islands (for ignimbrite emplacement) followed by subsidence (for the deposition of marine sediments). They defined two terms for welded textures: **eutaxitic** (Fig. 3e), referring to the agglutinated and flattened glass shards and their distortion around crystal and rock fragments; and **parataxitic** (Fig. 3f) referring to the extreme flattening and stretching of the original glass shards. Both terms are now commonly used. Many of the welded tuffs also contain dense dark coloured collapsed pumice fragments and these are termed **fiamme** (Fig. 4).

Rast (1969) and Bromley (1969) extended the concept of subaerial volcanicity in North Wales by proposing that a central volcano in Snowdonia dominated the Caradocian (mid Ordovician) period of volcanicity. They claimed to recognise a **caldera** − a large circular collapse structure which occurs at the summit of many volcanoes − and although

Figure 4 Acidic ash-flow tuff, welded, with chloritised fiamme drawn out along welding foliation (IGS photograph).

their analysis has not found general acceptance, their broad proposition of an extensive subaerial expression of the volcanicity in Snowdonia gained a great deal of support in the 1960s. Only recently has doubt been expressed, by geologists from the Institute of Geological Sciences working in northern Snowdonia, as to whether welding in ash-flow tuffs is diagnostic of a subaerial environment. This group has proposed that the tuffs can retain sufficient heat to weld under water, and their proposal has gained considerable support as a result of the very recent determination, by geologists from the University of Rhode Island, New York, of welding in ash-flows in a submarine environment off the island of Dominica in the West Indies. The proposal first made over 150 years ago, that most of the extrusive volcanic rocks of North Wales were emplaced in submarine environments, now appears much more plausible.

Careful mapping of the ancient volcanic sequence in Snowdonia indicates that there were several volcanic centres, which were active in different areas at different times. As a result, a great variety of volcanic deposits of varying composition and structure were formed. This fact will be more readily appreciated following some of the excursions outlined in this book, and it is to be hoped that the reader will be able to form his/her own opinion as to how the various volcanic rocks were emplaced.

Sedimentary rocks

Sedimentary rocks form a major part of the geological succession of Snowdonia, and an appreciation of their nature is essential in determining the geological evolution of the area during Lower Palaeozoic times. Ideally, interpretation of the sedimentary rocks involves detailed studies, but interested observers armed with a basic understanding of sedimentary processes can expect to reach some general conclusions on how the rocks were formed and on the conditions prevailing at that time. The rocks in Snowdonia have been changed by recrystallisation and grain movements, due to burial (**diagenesis**), and subsequently by heat and pressure during regional compression of the Earth's crust; but the degree of alteration is limited, and original sedimentary characters are still readily discernible.

Sedimentary rocks are consolidated sediments and materials that have segregated and accumulated, generally at low temperature and pressure, at the Earth's surface. They are characterised by a layering termed **bedding**, with each bed representing a distinct phase of sedimentation. Sedimentary, igneous and metamorphic rocks are the three main groupings that embrace all rocks. The sedimentary group includes **clastic rocks** (those composed of fragments), chemical precipitates (e.g. ironstone) and organic accumulations (of which coal is significant). The major clastic division comprises **pyroclastic** and **epiclastic rocks**. Epiclastic sediments form the bulk of all sedimentary rocks and are composed of grains derived from older rocks – grains that may be single minerals or rock fragments. They are classified in terms of grain size as: mudstones ($<1/_{256}$ mm), siltstones ($1/_{256} - 1/_{16}$ mm), sandstones ($1/_{16} - 4$ mm) and, depending on whether the fragments are rounded or angular, conglomerates and breccias (>4 mm).

Epiclastic sedimentary rocks predominate in Snowdonia. They consist of the fragmental detritus produced by the denudation of upstanding land areas by weathering and erosion. Weathered material is removed from source areas by water, either in solution in ground water or as discrete particles in surface streams. Where the streams have high energy, the smaller particles are held in suspension, while others, forming the bed load, are rolled or bounce along the bed of the stream. The abrasive effect of the bed load facilitates further erosion, and concomitantly fragments become smaller and more rounded downstream. As the stream loses its initial energy, a greater proportion of the sedimentary load is held in suspension, and the bed load moves along in migrating waves, small current ripples (Fig. 5) or larger dunes. When stream velocity decreases profoundly, energy is dissipated, and sedimentation ensues. Ideally, a pattern of deposition is established, with coarse particles being deposited before the fine. Loss of energy may occur in a variety of situations, such as where a river enters a large body of standing water (a lake or the sea), or

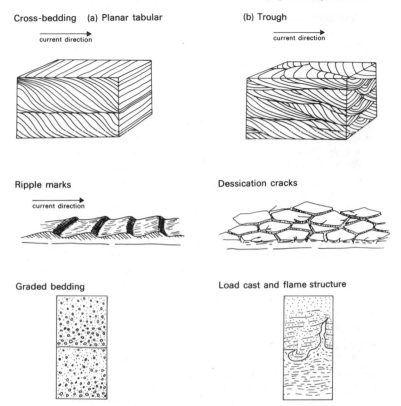

Figure 5 Diagram showing some of the sedimentary structures to be seen in Snowdonia.

breaches its banks or discharges from an area of high relief to an area of gentler topography.

In arid land areas quantities of accumulated detritus may be rapidly removed from high ground by temporary streams following heavy infrequent rains. When the streams emerge into an adjacent low-relief area (**piedmont**), they form either a sheetflood that spreads out radially or a radiated series of streams. Material is rapidly deposited, with coarse gravels near the point of discharge passing out into thinner deposits of sand and silt. The deposit, semi-circular in plan, is termed an **alluvial fan,** and an accumulation of successive beds makes an **alluvial cone** (Fig. 19). With greater rainfall, more permanent streams may transport detritus across the fan on to adjacent flat-lying areas; there they spread to form an interconnecting system of smaller streams (**braided streams**), which deposit

fluvial sediments. Some of the lowest Cambrian rocks and the coarsest sediments in the Ordovician are alluvial fan conglomerates and related fluvial sandstones. However, more common than the land-formed **(terrestrial)** sediments are the deposits formed in the transitional zone between land and sea and in a marine environment.

The reduction of the rate of flow of a river on entering the sea is less rapid than at an alluvial fan, but a similar process of flow dissipation and sedimentation occurs. The type of deposit formed depends on many variables, such as the continuity and rate of supply of sediment, depth of water, and the effectiveness of marine wave, tidal and current action in redistributing deposited material. Amounts of sediment transported by rivers can be considerable, the Mississippi, for example, moving some 2 000 000 tonnes per day; and where the amount of material deposited by the river exceeds that removed by marine action, **deltas** develop. They form by the thick accumulation of coarse material at the mouth of the river, with the fine sediment, carried further out into the sea, giving thinner deposits. The inclined surface of the deposit is called the **delta front**, and this advances when further layers, termed **foreset beds,** are deposited. Many of the major deltas of today are long established, with deltaic deposits thousands of metres thick, much deeper than the surrounding sea. For this to have happened there must have been continuous sediment supply accompanied by subsidence in the delta area.

A similar type of sedimentation is recognised where streams flow from areas of low relief into shallow water. A small-scale shallow delta builds up to sea level, above which deposition cannot proceed, and the front rapidly advances, forming a sheet deposit. This may be either attacked by wave action, if sediment supply ceases, or inundated and covered with marine deposits. Clearly, the process may then be repeated, with interdigitated fluvial and marine deposits, known as **paralic sediments**, being developed.

The supply of sediment and available space in the depositional area are critical in determining the character of a sedimentary sequence. Earth movements, which may cause uplift or subsidence of the Earth's surface, are the controlling factor. Major movements result in transgression of the sea on to the land or regression of the sea from the land. Uplift of a source area produces vigorous erosion with an abundant supply of sedimentary materials, and associated subsidence in the depositional area gives space for thick accumulations to be deposited.

Such earth movements, perhaps accompanied by earthquakes, commonly disturb the rapidly deposited near-shore sediments, which become unstable, collapse and move downslope as **turbidity currents** — turbulent currents carrying coarse and fine sediment, flowing along the sea floor and maintaining their identity. As the flows lose their momentum, sediment settles to form distinctive deposits known as **turbidites**, which characteristically are poorly sorted, containing a large range of particle sizes. Although turbidites may form in shallow water, the turbidity current is one way by which coarser sediments may reach the deeper water of the bathyal environment, where normal sedimentation is of fine mud carried out to sea in suspension. In 1929 an earthquake off Newfoundland triggered a turbidity current, whose rate of flow, indicated by the timed fracturing of submarine cables, exceeded 100 km/h. Its extent was subsequently shown to be several thousand square kilometres.

In contrast are the deposits of the shallow-water marine environment, called the **neritic zone.** Here, where sediment supply is limited, deposits may be attacked by wave and tidal action and sorted, the fine particles being removed by currents and the coarser sand grains remaining. Quartz grains, which are chemically and mechanically stable, are concentrated in these conditions, and the lithified deposits are termed **quartzites.** With a more abundant sediment supply tidal flats, backed by beaches of winnowed sand, and lagoonal areas may develop. In the intertidal environment, called the **littoral zone,** silt and mud alternating with thin wispy layers of sand sorted by tidal currents are characteristic. These may be disturbed by the animal life that abounds in this setting. The uppermost Cambrian and basal Ordovician rocks of our area are such shallow-water deposits.

Epiclastic rocks, both coarse and fine, can thus occur in a variety of environments, both terrestrial and marine; but the setting and process of emplacement leave an imprint on the rocks, which can be observed and interpreted. As there is environmental control of sedimentation, rock type **(lithology)** and sedimentological and palaeontological features, all of which constitute the rock **facies,** characterise a sediment as having formed in a particular environment. Important in analysis are lithology, provenance, grain size, sorting, roundness of grains, and sedimentary structures.

Sedimentary structures that can be observed in the field are essentially bedding characters formed at the time of deposition and as a result may indicate mode and in some cases environment of deposition. The external shape of beds varies from extensive sheets to lenses and narrow 'shoestring' bodies. Thicknesses can vary from laminations (<1 cm) through thin to thickly-bedded massive units (>120 cm). Common internal structures are **cross-bedding** (Figs 5, 6) − also called **cross-lamination, false bedding** or **current bedding** − and **graded bedding** (Fig. 5). Cross-beds are minor layers oblique to major bedding surfaces, which indicate the current flow direction at the time of deposition. Graded beds display an upward decrease in maximum grain size and are a feature of turbidites. **Contorted** or **slumped beds** (Figs 6, 24 and 38) are also commonly associated with turbidites and result from the downslope movement of unconsolidated sediment. Contorted bedding may also result from the unequal loading and differential dewatering of unconsolidated sediment caused by the rapid

Figure 6   Bedding structures in sandstones and siltstones (locality 2, Excursion 2), even bedding in siltstones (A), cross-bedding in fine sandstone (B), and disturbed (slumped) bedding in fine sandstone (C) (IGS photograph).

deposition of the layers above. Downwarps of the overlying bed are **load casts** (Fig. 5), and intrusions of underlying sediment into the bed above are **flame structures** (Fig. 5). **Ripple marks** (Fig. 5), which form when a current flows over a sand surface, are common in fluvial or shallow marine sands. These are the small waves of sand that generally form at right angles to current flow and can be seen on any modern flat beach.

No one character of the epiclastic sediments is diagnostic of emplacement regime, but considered together they can be distinctive and on this basis the sedimentary rocks of northern Snowdonia have been interpreted. The conclusions are fundamental to the reconstruction of the geological history of the area.

Tectonic structures

Following deposition and consolidation, the Lower Palaeozoic rocks of Snowdonia were deformed into a major downfold, referred to as the **Snowdon Synclinorium** (a complex synclinal structure composed of a number of smaller folds), during a period of earth movements known as the **Caledonian orogeny**. Features of the rocks resulting from the deformation are known as **tectonic** structures, the best-known examples being the folds that result from the bending of the originally horizontal strata.

Folds that are convex upwards are **anticlines**; those that are convex downwards are **synclines**. Open folds have a wide angle between the sloping sides or limbs; whereas if the angle is small, the fold is tight. A **fold axis** is the line joining the points of maximum curvature on a folded bedding surface (i.e. the line about which the bed appears to have been bent), and an **axial plane** is the plane containing the fold axes of successive beds within the fold. Large scale folds in Snowdonia, such as the Idwal Syncline (Fig. 22) and Tryfan Anticline, are well displayed in partial section in the steeply dissected terrain (Excursions 5 and 6). Smaller scale folds, seen within single outcrops, are common in the Cambrian rocks, but not all minor folds are the result of tectonic deformation; some are the product of sedimentary processes (Fig. 24) or of flow in lavas (Fig. 40). Discrimination of the different types of folds is important if the history of the rocks is to be fully understood.

Other tectonic structures form during folding. Cleavage is ubiquitous in the rocks of North Wales, and it is their particularly well-developed cleavage that has made the slates of the area world famous. Cleavage forms in response to compression of the rock body and involves the physical rotation and recrystallisation of platy minerals, such as the micas, in parallel planar orientation. The cleavage planes form approximately at right angles to the main direction of compression and parallel to the axial planes of folds formed during the same compressional episode. The cleavage in the rocks of Snowdonia generally trends northeastwards and dips steeply, indicating that the principal compression during the deformation of the rocks was approximately horizontal and directed between the northwest and southeast. It is developed most conspicuously in fine-grained sedimentary rocks, such as siltstones and mudstones, where it is often the dominant structure, the original bedding being difficult to distinguish (Fig. 7). In these finer rocks the whole fabric has undergone reorganisation during stress, most grains having moved or recrystallised to some degree, giving a very pervasive cleavage termed **slaty cleavage**. In the coarser-grained rocks cleavage is often less well-developed and cleavage planes are more widely spaced, being formed partly by brittle fracture (breaking of the rock) rather than by a wholesale reorganisation of the

Figure 7   Cleavage–bedding relationship in interbedded sandstone (S) and siltstone (Si), cleavage being better developed in the siltstone bands. Note the slight refraction of the cleavage at the siltstone/sandstone contact.

mineral fabric. This latter cleavage, sometimes called **fracture cleavage**, is more correctly regarded as a type of joint development, joints being the fractures that form when a rock breaks under stress. During the formation of folds in strata of contrasting lithologies, cleavage may be formed preferentially in some layers and joints may predominate in others. The geometric relationships of the structures is usually orderly and predictable, and users of this guide will be able to examine these relationships at a number of localities (e.g. locality 4, Excursion 2).

When a volume of rock is compressed, it tends to deform (change shape), as can be seen when strata are folded. A body of rock subjected to compression will tend to become shorter in a direction parallel to the principal direction of compression and, if the volume does not change, will expand in directions perpendicular to that direction. If we can measure the shape of deformed objects in the rock body, such as fossils whose shapes before deformation are known, we can calculate the change in shape of the rock body, assuming that the deformation of the object is more or less equal to that of the rock. A structural geologist, Denis Wood, has shown that during deformation the rock body of the Cambrian slates of Snowdonia was shortened by as much as 50% in a horizontal direction and thickened vertically up to twice its original sedimentary thickness (Wood, 1974). These conclusions were reached from a study of the shapes of green spots, known as **reduction spots**, that occur in the slates. The spots, formed by migration of iron soon after the sediments were deposited, were originally spherical. Flattening during tectonic deformation of the rocks changed them to ellipsoids (Fig. 8), whose shapes can be measured and used to calculate the amount of deformation in the rocks. Maximum extension due to flattening is found to be in the plane of the cleavage. The deformed spots can be observed in many of the slate quarries; there are particularly good examples in Alexandra Quarry near Moel Tryfan (Excursion 2), where those interested can evaluate for themselves the change in shape of the rock body that took place during the folding and the formation of the slaty cleavage. **Faults** (large scale fractures across which the strata are displaced) are widely developed in Snowdonia. Many of the faults formed during the late Silurian and Devonian earth movements, some of the larger reflecting the presence of pre-existing fractures in the

Figure 8 Deformed reduction spots on a joint face in slate (IGS photograph).

older rocks beneath the Welsh Basin (p. 29). There has been later movement along some of the fractures, and a small earthquake was recorded as late as this century along the Aber-Dinlle Fault near the Menai Straits. The movement along faults also played an important part in the early evolution of the Welsh Basin, particularly at times of enhanced volcanic activity. Geological mapping shows that the development of contemporaneous faults locally controlled the disposition and lateral extent of both sediments and lava flows. Good examples of these early faults can be seen on the southern side of the Llanberis Pass (Excursion 8). Regional structures, such as the Snowdon Synclinorium may also have started to develop at a relatively early stage as their disposition affected Ordovician sedimentation.

Glaciation

Glaciation played a major role in shaping the landscape of Snowdonia. Ice was responsible for the erosion of vast amounts of rock debris, its removal and subsequent deposition, each process resulting in a variety of typical glacial landforms.

Ice is an efficient agent of erosion only when it is thick and flows reasonably quickly in a concentrated stream. Acting on its own, it has very little erosional strength and cannot tear solid rock away from its foundations or significantly abrade it. If, however, the rock surface has previously been broken up by the expansion of freezing moisture in cracks and joints in the rock, the ice can easily incorporate the loosened rock debris in its basal layers and carry it away. When this happens, the included rock fragments at the base of the moving ice become powerful erosive tools, scouring the surfaces over which they travel and prising loose still more blocks.

Glaciated valleys have a characteristic U-shaped cross-section (Fig. 19), which reflects the broad downcutting base of the moving glacier and contrasts with the V-shaped cross-section of a stream-cut valley. The longitudinal profile may undulate and contain abrupt downward steps and low ridge-like **rock bars** extending across the valley (Excursion 4). Commonly, shallow depressions (rock basins) occur in the valley floor between successive rock bars, and in these lake water is often impounded (Excursion 4).

The typically undulating longitudinal profile of a glaciated valley reflects irregular erosion caused by a number of factors. Clearly, a major factor is the varying 'strength' of the rocks in the path of the ice; it is fairly obvious that hard massive rocks will be more resistant to erosion than softer or highly cleaved and jointed rocks. The presence of cleavage or joint planes in the rock is of particular importance, as the expansion of moisture freezing in these planes will fragment the superficial rock and render it even more susceptible to erosion. Also, as most valley glaciers occupy pre-existing steam-eroded valleys, original irregularities in the valley floors may affect the movement and hence the erosive power of the glacier at various places along the valley floor. A further cause of downcutting is the variation in the volume of ice in a valley.

In addition to excavating deep valleys, glaciers may also carve out large amphitheatre-shaped recesses in the mountain sides, known as **cirques**. Typically, de-glaciated cirques consist of precipitous crags on three sides around a roughly circular rock basin, perhaps containing a lake, with a low outlet over a rocky lip on the fourth side. In Wales, cirques are known as **cwms**, and excellent examples abound in Snowdonia.

Many of the largest cwms, such as Cwm Idwal (Excursion 5), formed initially during the passage over the mountains of a thick ice sheet, and mark places where the ice began to downcut rapidly. During the late glacial stage these cwms became the sources of valley glaciers and were further enlarged, while many smaller cwms with all the characteristics of their larger neighbours were developed predominantly on northeast-facing slopes.

Locally, the direction of ice movement in glaciated areas can be deduced by examination of the rock surfaces. Surfaces formerly beneath glaciers commonly show scratches **(glacial striations)** caused by angular rock fragments dragged across the surface by the moving ice. The striations lie

Figure 9 Roche moutonnée, at the head of Nant Ffrancon, in sandstones underlying the Pitts Head Tuff. Glacial striations parallel to the geologist's hammer (IGS photograph).

parallel to the direction of ice movement, and sets of striations that cross one another indicate a change in the direction of flow. Outcrops scoured in this way often occur in elongate rounded hillocks known as **roches moutonnées** (Fig. 9) — a term introduced in 1787 by H. B. de Saussure, who, noting that they often occurred in groups, likened the topography that they produced to a wavy style of wig, or *moutonnée*, that was popular in his day.

The erosional effects of glaciers result in the removal of huge quantities of rock debris, transported by moving ice and melt water to be deposited elsewhere. Such deposits, traditionally known as **glacial drift**, are of two types: deposits that are layered, and those which are not.

The layered deposits commonly consist of alternating coarse and fine sand and gravel deposited from streams released from melting ice. They are referred to as **fluvioglacial deposits** (Fig. 14) and include great aprons of sand and gravel laid down in front of melting glaciers as well as deposits laid down from water flowing beneath the ice.

Unstratified drift (**glacial till**) is normally deposited directly from melting ice without an intervening stage of transport by flowing melt water. It characteristically comprises large boulders and cobbles randomly scattered through a matrix of much finer sand, silt and clay. Till deposits can be very thick and extensive, spreading like a blanket over the previously ice-covered terrain. On geological maps in Great Britain published by the Institute of Geological Sciences, this type of till is called **boulder clay** — a term that reflects its constituents. In North Wales two types of boulder clay have been recognised: one originating from land-based glaciation, the other originating from an Irish Sea ice sheet and containing many distinctive boulders brought from areas far beyond the borders of North Wales.

More local accumulations of till, deposited at the edges of glaciers in the form of distinct mounds or ridges, are known as **moraines** (Fig. 44). The movement of ice within a glacier acts like a conveyor belt, carrying both within the ice and on its surface the rock debris that it has scavenged during its passage. At the front and sides, as ice melts, the debris is dumped from the conveyor system. If the glacier snout is more or less stationary for any length of time, an elongate mound of rock debris extending all along the glacier front (a **terminal moraine**) results. Pauses in the retreat of a glacier caused by climatic change can result in the deposition of a series of terminal moraines at various points along a glacial valley. An advancing glacier cannot leave a similar record, as it will over-ride and destroy its terminal moraines as it extends downvalley. Mounds of debris deposited along the valley sides are termed **lateral moraines**. In Snowdonia, some of the finest moraines are sited in the cwms, where they mark the final retreat of the cwm glaciers about 10 000 years ago.

Cold climate processes affecting the land bordering areas of glaciation are called **periglacial processes**. The most important are related to cycles of freezing and thawing, and in Snowdonia these continue to operate to a minor extent during particularly severe winters of the present day.

Many of the extensive **screes** that have accumulated on the steep slopes below the crags in Snowdonia result from the freeze–thaw shattering of rocks in the cliffs above. Similarly, the process affects cleaved and jointed rocks in the flatter ground on the mountain tops and creates **blockfields**, consisting of great spreads of jumbled ice-heaved blocks. They are particularly well developed across the Carneddau and Glyders summits and may merge downslope into rock streams and screes.

Periglacial conditions commonly produce mass movement of superficial material previously deposited on sloping ground, and this creates distinctive landforms. These are developed when the surface layers of permanently frozen superficial deposits thaw and the water-saturated upper layer creeps downslope and over the still frozen subsurface layers. This process is called **solifluction**. If the downslope creep is halted by a change in climate, distinctive **solifluction lobes** or **terraces** may be preserved at the flow fronts.

Some interesting small-scale features known collectively as **patterned ground** develop under periglacial conditions; in Snowdonia they form to the present day. **Stone polygons** are one type of patterned ground that can

Figure 10 Patterned ground. Stone stripes and ill-formed polygons, near Foel Grach. The open compass is 15 cm long (IGS photograph).

be seen; they develop in rocky surface debris resting on flat ground at high altitudes, and consist of networks of polygonal shapes outlined by larger fragments with the central portions infilled by finer debris (Fig. 10). The polygons are generally a few tens of centimetres across. On steeper slopes **stone stripes** develop parallel to the hill slope; they comprise narrow alternating strips of larger and smaller fragments (Fig. 10). Both types of patterned ground result from the sorting and rearrangement of the debris during freeze–thaw cycles operating on moisture between the fragments. Stone polygons artificially disarranged during the autumn have in some cases been found naturally reassembled after the following winter – but please do not destroy any polygons that you find, as the process cannot be guaranteed!

3   Geological history

In recent years a new theory has been proposed that attempts to explain the distribution of earthquakes, volcanoes and tectonic activity on the Earth's surface. The theory proposes that the present surface of the Earth consists of six major **plates**, which are, to some degree, continuously moving with respect to each other. Most of the earthquakes, volcanoes and tectonic movements occur around the plate margins and relate to the nature of the contact between adjacent plates. Determination of the distribution and extent of the plates, and of the way in which they moved further back in geological time, is more difficult. However, it is thought that during the Lower Palaeozoic the great thicknesses of rock in Central and North Wales were deposited in a NE−SW trending basin situated within a plate of continental crust and close to a convergent plate margin where oceanic crust was being subducted beneath the continental plate.

The continental crust of older rocks, upon which the Lower Palaeozoic basin developed, is now only exposed in small outcrops at the edges of the basin: in Anglesey, Lleyn and the Bangor and Padarn ridges in the north, and in Pembrokeshire and the Welsh Borderland in the south. The relationships of these older rocks are problematic, although it is thought that the rocks of the Bangor and Padarn ridges are younger than those of Anglesey and Lleyn and may be early Cambrian in age.

On the Padarn ridge (Excursion 1) near Llanberis, the Precambrian or earliest Cambrian rocks comprise a thick sequence of acidic ash-flow tuffs, which are overlain by alluvial fan, fluvial and shallow water deposits, forming a sequence of conglomerates, sandstones and siltstones, with a few thin acidic ash-flow tuffs. The junction in places is apparently conformable, elsewhere discordant and locally faulted. The character of these lowest Cambrian sediments indicates an availability of coarse material from within the area of the ridge, suggesting continuous uplift in this area and rapid accumulation.

In contrast, the succeeding lower Cambrian sediments are dominated by the purple and blue−grey mudstones from which much of the famous slate of North Wales is derived. These clearly reflect more stable quiescent conditions, only interrupted occasionally by incursions of coarser debris, which spread as turbidity flows across the basin floor. Coarser turbiditic sandstones are the dominant lithology of the middle Cambrian; they reflect a reactivation of the sediment supply by uplift in the source area, but without contemporaneous volcanism. There was a return to more stable conditions in the upper Cambrian, when fine siltstones with thin interbedded sandstones were deposited. The uppermost sandstones, rich in quartz, are the products of marine reworking of older sediments in increasingly shallow-water conditions. Regional uplift, greatest towards the west and accompanied by erosion, took place in the late Cambrian and was followed by an early Ordovician (Arenig) marine transgression.

Shallow-water bioturbated sandstones form the basal Ordovician strata, which rest on successively older Cambrian horizons as the Padarn and Bangor ridges are approached. These are immediately succeeded by mudstones, siltstones and impersistent sandstones of a littoral environment, which pass up into neritic or bathyal siltstones and mudstones in which occasional temporary shallowing is indicated by lenticular beds of ironstone.

The fairly stable pattern of sedimentation was disturbed by tectonic activity within the basin, which caused the formation of slump breccias upon submarine slopes. This activity heralded the first major phase of volcanism in the area, expressed in the development of a series of volcanic centres, now recognised in the Llewelyn Volcanic Group (Excursion 3), which extends northwards from Nant Ffrancon through the Carneddau but is not represented in southern Snowdonia. These centres were volcanically distinctive, erupting acid, intermediate and basic lavas and pyroclastic rocks. Towards the end of this activity the associated sediments indicate uplift and emergence in the north of the area, with the widespread development of terrestrial alluvial-fan conglomerates in the north and shallow marine sandstones in the south. The volcanic episode terminated with the most extensive eruption: the acidic ash flows of the Capel Curig Volcanic Formation. The lowest two ash flows extended from a subaerial vent in the north into a marine environment in the south.

A period of relative volcanic calm followed these eruptions, and the broad sedimentary pattern prior to the activity was re-established, with coarser deltaic sandstone deposits to the north passing southeastwards into thicker offshore siltstones. Minor volcanic eruptions did occur, and these are represented by the widespread fine air-fall water-settled dust tuffs that break the sedimentary sequences.

In central and southern Snowdonia attenuated sequences of shallow water sandstones reflect uplift prior to the emplacement of the acidic ash-flow of the Pitts Head Tuff. This is assumed to have been erupted from a centre in southwestern Snowdonia and extended as far north as the Nant Ffrancon Pass. Following the emplacement of the Pitts Head Tuff, the predominant sandstone deposition continued with more frequent eruptions producing air-fall dust tuff (Excursion 5). In southern Snowdonia this pattern was broken by a major collapse of the sediments, probably caused by magmatic doming and represented by the Llyn Dinas Breccias. These breccias occur about Snowdon and are at their maximum development on the southern side. However, in the Llanberis Pass (Excursion 8) the uplift is represented by contemporaneous faulting. This tectonic activity heralded a period of intense volcanic activity with the eruption of the pyroclastic rocks and lavas of the Snowdon Volcanic Group. Through this period there is evidence that in central Snowdonia the volcanism had at least a temporary subaerial expression. However, in the surrounding area marine conditions prevailed. The earliest products of the activity are: the acidic ash-flows and rhyolites of the Lower Rhyolitic Tuff Formation, which formed an unbroken accumulation in central Snowdonia (Excursion 14); a succession of primary ash-flow and secondary slumped pyroclastic deposits in the Idwal area (Excursion 8); and in the north and east, three major acidic ash-flows, comprising the Lower Crafnant Volcanic Formation, which escaped into the adjacent marine sedimentary basin (Excursion 9), breaking the sedimentary sequence.

Following these earliest widespread acidic eruptions, a separate acidic volcanic centre developed in the deeper water of northeastern Snowdonia, while local basic volcanism, producing the Bedded Pyroclastic Formation, was established in fairly shallow marine conditions in central Snowdonia. Through this period the associated sediments in the area to the north and on the eastern margins of Snowdonia show a marked decrease in the supply of coarser material, and a black mudstone lithology is characteristic prior to the final widespread eruption, which formed the uppermost unit of the group: the Upper Rhyolitic Tuff Formation. This eruption was of ash-flow type, although during the course of its emplacement it incorporated a distinctive amount of the unlithified black mudstone.

The interpretation of the geological history following the Snowdon Volcanic Group has to be restricted to northeastern Snowdonia, as it is only here that the succeeding strata are exposed. In this area local deep-water basic volcanic centres (Excursion 13) developed in the deep-water black mudstone environment. They eventually waned as the sedimentary environment in the upper Ordovician became shallower, and the lithologies tended towards those of the succeeding Silurian System.

Apart from a small enclave on the western side of the Conwy Valley in the Conwy area, Silurian strata occur entirely to the east in the old county of Denbighshire (now part of Clwyd). However, it is clear that originally these rocks must have overlain the Ordovician across Snowdonia. The Silurian rocks of former Denbighshire are entirely free of volcanic influence, and the style of sedimentation is in sharp contrast to that of the Ordovician. Striped mudstones, turbiditic sandstones and slumped sequences reflect basinal sedimentation in a tectonically active environment. This suggests that the whole of the Ordovician and Cambrian sequence was buried in the rapidly subsiding basin.

This description of events occuring in the Lower Palaeozoic is based entirely on the interpretation of the lithological sequence, with no reference to the fossil content of the rocks. As many readers will be aware, in adjacent areas the systems have been divided into Series on the basis of the graptolites and shells that they contain. However, in northern Snowdonia fossils occur only sporadically, and many problems of faunal correlation remain to be resolved, especially with regard to the relationships of the graptolite zones to the shelly stages in the Ordovician. The lowermost Ordovician of the area is the Arenig Series, identified by the graptolites, whereas the main expression of the volcanism is assigned to the mid-Ordovician Caradoc Series, on the basis of shelly fossils. In contrast, the Silurian rocks have yielded much faunal evidence, and the graptolite zones are firmly established.

Following the deposition of the Silurian rocks, earth movements, which had continuously affected the basin and its deposits throughout the Lower Palaeozoic, increased in intensity, folding the rocks into the structures seen today and elevating the rocks of the basin into a huge mountain range. These earth movements are referred to as the **Caledonian orogeny** (see p. 108) and the mountain range as the **Caledonides**.

Following the development of the Caledonides, the geological history of Snowdonia is less well known until events in the Pleistocene. From the record of the Upper Palaeozoic and Mesozoic rocks in adjacent areas it is reasonable to assume that there were several periods of submergence with deposition of strata, followed by uplift and erosion, which in Snowdonia removed all traces of the record.

It is apparent from the evidence of offshore basins that the form of the British Isles as we know it today was established by the end of the Tertiary era. Towards the end of the era the climate changed from temperate to cool. The change continued, and during the Pleistocene period ice covered most of northern Britain at some time. The ice sheet in North Wales, known as the Merioneth Ice, was at least 650 m thick and was centred on an inland plateau to the east of Snowdonia. It extended westwards, over-riding the mountains of Snowdonia, so that only the highest peaks and ridges emerged, and it had a profound erosional effect on the underlying landscape. A further sheet of ice occupied the Irish Sea and moved southwestwards along the coast of North Wales. At a late stage in the Ice Age the main ice retreated to isolated centres, and Snowdonia appeared much as the Alps of today, with snow-covered peaks and with glaciers occupying many of the valleys. About 10 000 years ago the glaciers retreated to the innermost recesses of the high cwms before finally disappearing. During this stage severe frosts and meltwater streams from the decaying ice continued to play a role in shaping the scenery and in redistributing the debris previously transported by the ice sheets and valley glaciers.

Part 2: Field excursions

Introduction

The excursions are entirely in the Snowdonia National Park, within which most of the ground is owned privately or by the Forestry Commission. Users of these excursion itineraries are requested to **conform to the Code for Geological Field Work**, which is available on request from the Geologists' Association; a condensed version is to be found on the back cover of this book. Permission should be obtained for entry to any areas that are clearly not open sheep grazing or public footpaths. The Forestry Commission permits walkers to use its tracks, but motorists require permission for access. Clearly, in the afforested areas the commission has its own regulations, details of which will be provided on request from its office at Gwydir Uchaf, Llanrwst.

In the park, hammers should be used very sparingly; at most localities examination of the weathered surfaces will provide the greatest instruction. Where close examination of rock faces is necessary, a hand lens is a useful accessory. On all the excursions stout shoes or boots should be worn, and for the high ground excursions protective wear and spare clothing should be carried.

The area of the excursions is covered by the Ordnance Survey 1:50 000 Sheet 15 and by two of the 1:25 000 Ordnance Survey Outdoor Leisure Maps of the Snowdonia National Park: the Snowdon and Conwy Valley sheets. All grid references such as [543 625], refer to the 100 km National Grid square SH.

1 Llanberis area

Duration: ½ day.

This excursion is to examine the oldest rocks exposed in northern Snowdonia (the Padarn Tuff Formation) and the overlying sedimentary sequence of early to middle Cambrian age. The lower Cambrian slates, extensively quarried in the Llanberis area, are well exposed on the route, and the excursion ends appropriately at the Quarrying Museum in the Padarn Country Park near Llanberis. The distance to be covered is about 8 km.

The strata to be examined present problems that have been debated since the Lower Palaeozoic systems were first erected in the nineteenth century. In particular, the precise position of the boundary between Precambrian and Cambrian strata is a matter of controversy to the present day. The oldest fossils found in northern Snowdonia are trilobites characteristic of the uppermost part of the lower Cambrian. However, a considerable thickness of strata – including, in descending order, most of the Llanberis Slates Formation, the Fachwen Formation and the Padarn Tuff Formation – underlies the fossiliferous horizon and presents a problem in establishing the base of the Cambrian System. Traditionally, this base has been assumed to lie at the boundary between the Padarn Tuff Formation and the overlying tuffaceous conglomerates, sandstones and siltstones of the Fachwen Formation, and early researchers believed that this boundary represented a profound unconformity. More recently, the existence of an unconformity has been questioned, and the age of the volcanic rocks of the Padarn ridge, whether early Cambrian or Precambrian, remains to be resolved.

The route (Fig. 11) starts at Bryn Brâs Castle [543 625] near Llanrug and can be followed either on foot or by car. Coach parties should alight at Bryn Brâs Castle, where there is a suitable turning point for coaches, and arrange to be collected at the Llanberis Quarrying Museum car park [586 604].

Follow the road southeastwards through the arch at Bryn Brâs Castle and uphill for approximately 1 km to a cattle grid beyond which cars can be parked (1) (numbers printed in bold type refer to localities, which are shown within circles on the figures). Here, on the northern side of the road, a dolerite dyke 20 m thick, trending generally eastwards, intrudes the Padarn Tuff Formation. A thinner dolerite dyke can be examined on the southern side of the road. The dykes intrude welded acidic ash-flow tuffs, and the welding foliation dips at a low angle to the southeast. At the roadside, about 100 m beyond the cattle grid, the foliation dips at a low

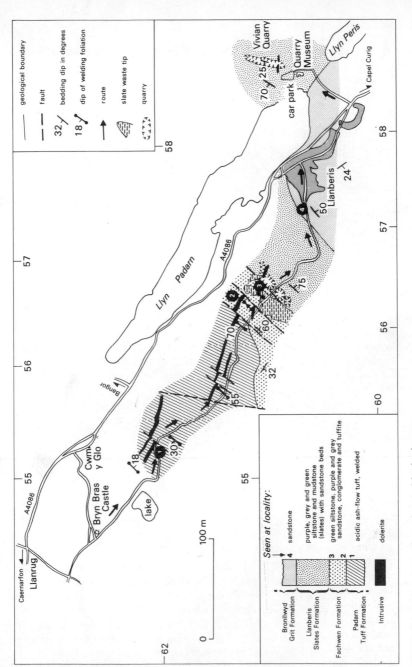

Figure 11 Sketch map. Excursion 1, Llanberis area.

angle to the northwest. Here the tuff consists of a fine-grained matrix studded with grey **euhedral** quartz crystals and pinkish flattened pumice fragments (fiamme) concordant with the welding foliation. The variation in orientation of the welding foliation indicates the presence of a small syncline, and its axial plane coincides with a fault that terminates both the dolerite dykes. These relationships should be examined across the outcrops.

The route continues eastwards along the road and crosses the Padarn Tuff Formation, with numerous ice-smoothed outcrops of the welded tuff forming the distinctive topography. Wherever the welding foliation can be distinguished, it dips at a moderate angle to the northwest. Across these outcrops there is little evidence of either air-fall tuffs, volcaniclastic material or sedimentary intercalations, which indicates that the activity producing the Padarn Tuff Formation was dominated by fairly continuous ash-flow eruption. The great thickness of the tuffs suggests that they were emplaced in a restricted depression, such as a fault-bounded valley, although there is no indication as to whether they accumulated in a subaerial or subaqueous environment.

From the road there are excellent views northwards over Llyn Padarn and southeastwards to the Llanberis Pass, with the glacially scoured terrain clearly displayed. The extent of the Llanberis Slates Formation, which lies to the southeast of, and stratigraphically above, the Padarn Tuff and Fachwen Formations, is clearly indicated by the numerous slate quarries and associated waste tips. Note that outcrops of the slates are much less evident than those of the tuffs, as the latter, being much harder, were more resistant to erosion by the ice that flowed northwestwards over the area during the Pleistocene period.

Continue eastwards along the road to a point about 50 m before it passes between two large slate tips. Outcrops at the roadside are of interbanded purple–grey and green–grey, coarse to medium-grained, feldspathic sandstone and siltstone of the Fachwen Formation. Follow a small footpath for about 150 m downhill to the large outcrops of conglomerate (2). These conglomerates form the lowest part of the Fachwen Formation and comprise abundant pebbles and cobbles of grey and pink quartz, quartzite and welded acidic tuff in a tuffaceous sandy matrix. Careful examination of these outcrops shows that the conglomerates pass laterally, to the south, into cross-bedded sandstones. The conglomerate has a lenticular form and was probably deposited from a fast-flowing stream as a relatively restricted alluvial cone. Much of the included debris, both clasts and matrix, was derived from the underlying Padarn Tuff Formation; but pebbles of older rocks, now only exposed on Lleyn and Anglesey, are also present. The relationship of the conglomerate to the underlying tuff is variable; in places the junction is conformable, even gradational, but elsewhere it is unconformable or faulted. At this locality it is faulted and lies to the west on private land, which should not be entered without permission. However, the faulted boundary can be inferred, as the conglomerates and sandstones dip to the northwest as if passing beneath the nearest outcrops of tuff, although the latter are in fact older.

Return along the footpath to the road, noting the acid tuff exposures in the Fachwen Formation, which indicate that volcanism continued for some time following the emplacement of the Padarn Tuff Formation.

Continue along the road between the slate tips, to the northwestern boundary fence of the slate quarry on the northern side of the road. Follow it downhill for about 150 m, passing exposures of sandstone with finer tuffitic siltstone bands in the upper part of the Fachwen Formation. Where the edge of the tip bends to the north (3), an outcrop exposes an infilled channel of coarse white sandstone in banded fine sandstones and siltstones. The sequence is folded and cleaved. Close examination shows that some of the thin siltstone beds protrude into the sandstone of the channel. This indicates that the channel was cut in a number of episodes separated by the deposition of silts and muds, the protruding siltstone representing part of the floor of each successive channel. The sandstone deposited in the channel is well exposed in a small cliff about 20 m to the northeast, where elongate rafts of siltstone, torn from the bottom and sides of the channel, are contained within the sandstones. To the west of this outcrop the underlying finer sandstones include light-coloured tuffitic siltstone bands, which reflect a continuing but waning volcanism during lower Cambrian times.

Retrace the route by the quarry fence to the road, and note the impressive views of the quarries in the Llanberis Slates Formation.The contact between the Fachwen and Llanberis Slates Formations lies approximately along the northwestern edges of the quarries and is here faulted. Although the predominant structure seen in the purple siltstone slates is a near-vertical 'slaty' cleavage, the bedding is apparent where thin bands of green sandstone, a few centimetres thick, are interbedded. Even from our roadside viewpoint careful observation of this bedding will reveal minor folds and faults on several of the quarry faces. Further along the road, just beyond the quarries, at a road junction on the right, sandstones forming part of the Llanberis Slates Formation can be examined.

Continue along the road towards Llanberis to the prominent outcrops north of the road at the edge of Llanberis village [572 603](4). The outcrops are of sandstone, part of the Bronllwyd Grit, which overlies the Llanberis Slates Formation. The sandstones occur in moderately thick beds, some of which display an upward grading of coarse to fine clastic grains indicative of deposition from turbidity flows. The Bronllwyd Grit spans the boundary between the lower and middle Cambrian, and its base marks the upper limit of the workable slates of the major Bethesda—Llanberis—Nantlle slate belt. Further outcrops of the Bronllwyd Grit can be seen between the houses along the road down towards Llanberis High Street.

The final localities of the excursion — the Quarrying Museum and Vivian Slate Quarry [586 604] — lie at the eastern end of Llyn Padarn, reached by a signposted road leaving the A4086 trunk road at the edge of Llanberis village. The quarry forms a part of the Padarn Country Park and can be entered from the museum car park or viewed from a higher level from one of the marked trails that ascend its northern side. The scale of these manmade excavations is most impressive when they are viewed from the quarry floor. Bedding and cleavage relationships in the slates can be examined in the faces. A dolerite dyke forms a prominent wall-like feature partially crossing the base of the quarry. The dolerite, being very hard, was avoided by the quarry men during excavation of the slate. From the path in the quarry floor the dyke can be examined in cross-section. The dark green rock is crossed by near-horizontal quartz veins, like the rungs of a ladder ascending the near-vertical dyke. The veins infill tension gashes produced by stretching deformation of the dyke.

After leaving the quarry, a visit to the nearby Quarrying Museum provides a fascinating insight into the history of the local slate-mining industry. Refreshments, toilets and attractive lakeside picnic sites are available around the car park of the Padarn Railway and Quarrying Museum complex.

2 Alexandra Slate Quarry

Duration: ½ day.

This excursion is to examine a classic glacial drift section and the slates and associated sandstones of the Llanberis Slates Formation. In the Alexandra Quarry [518 561] various tectonic and sedimentary structures are well displayed, and the degree of deformation of the slate can be gauged from an examination of deformed reduction spots.

The quarry (Figs 12, 13), now disused, lies to the southeast of the villages of Rhosgadfan and Pen-y-fridd, reached by turning southeastwards along a minor road from the Caernarfon–Porthmadog trunk road (A487), 1 km south of Bontnewydd. From Pen-y-fridd the northern end of the quarry is reached by a minor road and unpaved track. Coaches should turn at the beginning of the track; the quarry is then reached on foot – a distance of about 1 km.

The large waste tips around the northeastern end of the quarry provide a vantage point for fine views to the northeast and east, of Moel Eilio, the Snowdon massif and Mynydd Mawr, the latter comprising a large granite boss. The outcrop of the Llanberis Slates Formation is marked by the scattered small slate quarries extending to the skyline north of Moel Eilio, and on the western slopes of Moel Eilio the long narrow cleft of the old ironstone workings of Betws Garmon is visible.

Follow the track over the tips to the point where it divides, one branch leading down into the quarry, another to a group of old quarry buildings. From the old buildings walk southwestwards to the eastern rim of the quarry, where, between the tips and the quarry edge, boulder clay and fluvioglacial sands are exposed (1). The laminated sands contain cemented slabs with well-preserved flute-mark casts on their bases (Fig. 14) and occasional small fragments of marine shells. The outcrop is small and should not be disturbed; formerly, more sand was exposed in the area now covered by the tips. The discovery of the marine shells, early in the nineteenth century, led to a lively controversy. Proponents of the theory that all glacial drift was deposited from floating ice, following a major rise in sea level and a widespread inundation of the land, found strong support in the occurrence of the shell-bearing water-laid deposits at a height of over 300 m above the present sea level. It was subsequently realised, however, that there was an alternative and more plausible explanation. The sand and shells were originally incorporated in the base of the Irish Sea ice sheet, which at its greatest development extended inland as far as the Moel Tryfan area. When the ice eventually melted, the sand and shells collected from the

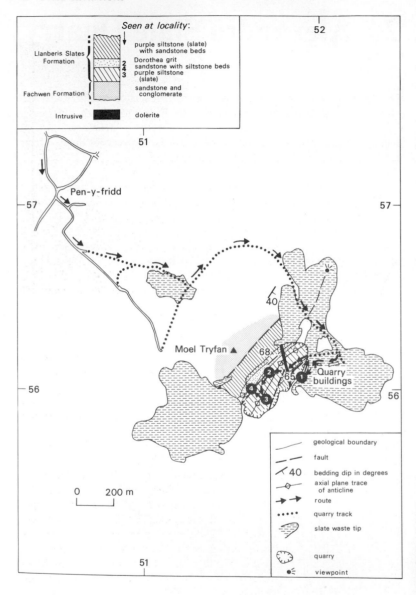

Figure 12 Sketch map. Excursion 2, Alexandra Quarry.

Figure 13 Annotated photograph showing route in Alexandra Quarry (Excursion 2) (IGS photograph).

sea floor and transported inland within the ice sheet were redeposited in their present position from meltwater streams flowing along the edge of, or beneath, the wasting mass of ice.

Descend the track into the northeastern end of the quarry, where a good general view of the excavation is obtained (Fig. 13). Slate-quarrying dominated local life for almost 200 years, when many of the surrounding villages, with distinctively biblical names, were established. The height of slate extraction was during the nineteenth century, to meet the demand of the building industry during the development of the industrial Midlands.

The quarry exposes a complex faulted anticline, which can be verified by observing the direction of bedding in various parts of the quarry. Purple cleaved siltstones are the dominant lithology exposed in the quarry faces, but along the western and part of the eastern side the excavation exposes the Dorothea Grit – an important sandstone horizon within the Llanberis Slates Formation. To the west of the quarry the rounded hill of Moel Tryfan consists of sandstones and conglomerates of the underlying Fachwen Formation, faulted against the Llanberis Slates Formation.

The Dorothea Grit lithology can be examined in the large blocks at the track side just beyond the flooded part of the quarry floor (2). It comprises graded coarse-grained sandstone bands intercalated with banded siltstone

Figure 14   Fluvioglacial sand and gravel. Blocks of cemented sand showing infilled flute mark casts on their undersurfaces (**1**, Excursion 2) (IGS photograph).

and cross-bedded fine-grained sandstone. The graded beds were deposited from turbidity flows, which spread over the silt and mud characteristic of the normally placid depositional environment of the Llanberis Slates Formation. Angular fragments of mudstone and siltstone were incorporated into the turbidity flow, in a partially lithified state, during its transport. In places the finer sediments have been contorted by slumping, and some of the silty interbeds between the sandstones are tuffaceous. It is possible that earthquake shocks associated with distant eruptions, which supplied the volcanic dust incorporated in the tuffaceous sediments, also triggered off the turbidity flows from which the graded sandstone units were deposited.

In the purple slates towards the southeastern end of the quarry (**3**) deformed green reduction spots up to 3 cm long are conspicuous on joint and cleavage planes. On cleavage planes the spots show only a slight tendency to preferential elongation, but on joints, perpendicular to the cleavage, the spots are markedly extended in the direction of cleavage and flattened in a direction normal to the cleavage. A quantitative estimate of deformation can be made by measuring the principal axes of the ellipses, calculating the volume of the ellipsoid and comparing the length of the axes with the diameter of a sphere of identical volume.

In the southwestern corner of the quarry (**4**) an anticline within the Dorothea Grit is seen in section. Cleavage is developed in the fine sandstone beds, but the coarser beds are cut mainly by joints, which are normal to bedding surfaces and hence fan around the anticline. The joints formed while the sandstone beds were folded. Cleavage is best developed in the siltstones to the west of the folded sandstone. The cleavage lies approximately parallel to the axial plane of the fold. The siltstones have been faulted against the sandstones in the western limb of the fold.

Return to the starting point, either by retracing the route or by leaving the quarry at its southwestern end and turning northwards over the summit of Moel Tryfan, where there is a good outcrop of conglomerate of the Fachwen Formation.

3 Ysgolion Duon

Duration: 1 day.

This excursion is to examine acid, intermediate and basic volcanic rocks and the associated sedimentary rocks of the lower part of the Llewelyn Volcanic Group – the first major development of Ordovician volcanics in northern Snowdonia. In addition, fine glacial landforms are displayed along the route. The first part of the excursion involves an easy walk along the secluded valley of the Afon Llafar to the spectacular cliffs of Ysgolion Duon (the Black Ladders). The second part is more rigorous, including the ascent of the ridge between the two highest Carneddau peaks; it should only be attempted in good weather conditions and with appropriate clothing and footwear.

The route (Figs 15, 16) commences at the southeastern end of the village of Gerlan [633 663], where there is limited roadside parking for cars and a turning point for coaches.

Walk along the minor road southeastwards from the village towards Ty Slatters Farm, turning right to cross the Afon Llafar, and reach the footpath striking southeastwards from a stile at the gate to the waterworks [638 659]. The path ascends gradually across thick boulder clay and eventually enters the Llafar Valley.

A little beyond a small iron-fenced enclosure and a small spur [654 650] the crags of Ysgolion Duon, at the head of the valley, come into view, and from this point (1) some of the prominent glacial features of the valley can be considered. Most conspicuous is the large moraine on the northeastern side of the valley, with its gentler sloping upper surface abutting against the steep flank of Yr Elen and extending into the cwms at the head of the valley. The large boulders comprising the moraine are exposed on the steep slope along the northeastern side of the Afon Llafar. This impressive moraine was deposited from the main Llafar Valley glacier. At a late stage a subsidiary channel, now occupied by the Afon Llafar, was cut along the western side of the main valley floor, the erosive agent probably being meltwater from the receding ice. Higher up the valley (see 4) small terminal moraines within this subsidiary channel indicate that during the late glacial stage it was partly occupied by ice emerging from the high cwms on the sheltered northeast-facing side of the main valley.

The route continues up the valley, with cleaved siltstones and mudstones of the lower part of the Ordovician sequence (Llanvirn and Llandeilo) cropping out on the slopes above the footpath. At (2) (Fig. 15), by a small waterfall [663 642], an outcrop of cleaved siltstone has been prospected for

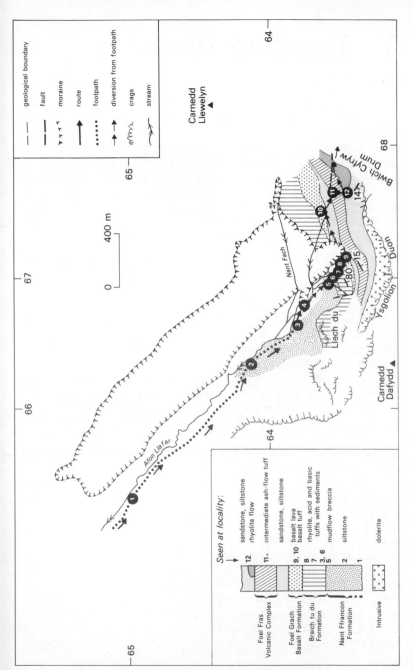

Figure 15 Sketch map. Excursion 3, Ysgolion Duon.

Figure 16   Annotated photograph showing route, Ysgolion Duon (Excursion 3).

slate. The siltstones have been thermally metamorphosed by nearby minor intrusions, and small, dark grey spots of chlorite can be seen on cleavage surfaces. The spots are concentrated in bands parallel to the bedding, and careful examination with a hand lens reveals that they have been deformed into ellipsoids. Their deformation was a response to the same stresses that formed the cleavage, indicating that emplacement of the igneous intrusions was earlier than the main episode of deformation in the rocks. From (**2**), note the view downstream, where the asymmetric cross-section of the Llafar Valley is clearly displayed, with the older moraine to the north and the later, more deeply cut, glacial channel to the south.

Continuing up the valley to (**3**) (Figs 15, 16), the path passes between massive boulders fallen from the crags above. The largest of these boulders (Fig. 17) is composed of rhyolite breccia, formed by the breaking of the chilled solid outer surface of a rhyolitic lava flow into blocks and their incorporation into the still hot and fluid core of the flow. Other large boulders near the path are of nodular rhyolite, some with incorporated blocks of flow-banded rhyolite. The rhyolites are from the lower part of the Llewelyn Volcanic Group and crop out in the nearby crags of Llech du.

The path continues close to the stream, crossing a small well-featured terminal moraine (**4**) in front of a flat marshy area once occupied by a small

Figure 17 Block of autobrecciated rhyolite (**3**, Excursion 3).

lake. Lateral moraines partially border the marshy flat, the moraines being deposited during a pause in the retreat of a glacier that once occupied the upper part of the deeply incised southwestern side of the Llafar Valley. The older spread of morainic debris occupying the major part of the Llafar Valley towers above these small late-stage moraines, and their age relations are here clearly displayed.

At (**5**), at the first outcrops slightly beyond, and to the south of, the old lake, a **mudflow breccia** is exposed. This consists of blocks and smaller angular fragments of vesicular basalt, acid volcanic rock, mudstone and siltstone in a dark grey mudstone matrix. The rock originated as a submarine landslip or turbidity flow, probably activated by volcano-tectonic activity, and its content of volcanic debris indicates that volcanism had already commenced in an adjacent part of the sedimentary basin. Similar breccias are consistently developed immediately below the lowest units of the Llewelyn Volcanic Group; they are the first indicators of major volcanism in the Ordovician sequence of this part of Snowdonia.

At a small waterfall (**6**) about 20 m upstream, the mudstone breccia is overlain by volcaniclastic sandstone containing coarse pebble layers, followed by a bedded acid tuff with distinctive lenticular cavities probably once occupied by pumice fragments. Just above the acid tuff (**7**) are much more massive basic lapilli tuffs, overlain by almost vertically dipping bedded tuffaceous siltstones; these are in turn overlain by an acid tuff similar to that seen at **(6)**. The same sequence crops out southwest of the old lake, but here it has been complicated by penecontemporaneous faulting and slumping. All these beds belong to the Braich ty du Volcanic Formation and display the heterogeneous nature of the early volcanism, with the influence of both acid and basic eruptive volcanic centres. This heterogeneity is repeated in both space and time throughout the Llewelyn Volcanic Group until the eruption of the uppermost units, the widespread ash-flow tuffs of the Capel Curig Volcanic Formation (see Excursions 6 and 7).

The nearby crags **(8)** to the southeast are formed of a rhyolite lava flow and should be ascended close to their contact with the boulder-strewn moraine. The rhyolite has poorly developed flow bands, which in the lowest outcrop dip steeply down the valley. The top of the flow is formed of about 2 m of brecciated rhyolite overlain by sediments. Examination of the contact shows that there was very little reworking of the flow surface, with siltstones, tuffaceous at the base, deposited about the rhyolite blocks. The succeeding sediments include several beds of breccia containing rhyolite and mudstone fragments. These beds were probably deposited from mudflows which, initiated at the rhyolite flow front, incorporated collapsed blocks and slabs of the rhyolite.

Continue close to the edge of the moraine. Outcrops of interbedded siltstone and breccia occur, although well-bedded cleaved siltstone and sandstone are more common higher in the sequence. In the outcrop **(9)** close to the top of the moraine, a basic tuff overlies the sediments. The tuff, in its basal 2–3 m, includes fragments of basalt with chlorite-filled vesicles, but in its upper part it is finer and bedded. This tuff is the local representative of the Foel Grach Basalt Formation and is the lateral equivalent of the basalt flow to be examined at **(10)**.

The excursion can be terminated at **9** should deteriorating weather conditions ᴄ .ack of suitable clothing dictate. If not, traverse eastwards at about 100 ᴵ . below the crags to the south, passing a small sheepfold, to the prominent basalt outcrops **(10)** north of a small stream. The blocky character of the basalt suggests proximity to the flow front, and this flow is possibly the source of the basic tuff at **(9)**. Both the basalt and basic tuff are part of the Foel Grach Basalt Formation.

At the base of the crags, south of the stream, a small dolerite dyke intrudes sandstones, which locally contain basaltic fragments. The sandstones are overlain **(11)** by a coarse greenish tuff breccia, rich in feldspar crystals and lithic fragments, interpreted as the basal zone of an ash-flow tuff of intermediate composition. Above, the tuff is massive and less obviously fragmentary and, on fresh surfaces, has the appearance of a lava rich in feldpar crystals. About 60 m above the base the tuff is bedded, representing the reworking of the top of the submarine ash-flow by wave action. This tuff is the local representative of the Foel Fras Volcanic Complex, which is at its thickest development to the north, about Foel Fras.

The tuff is overlain (12) by the brecciated west-facing flow front of a rhyolite lava. The brecciation can be traced from the base of the flow around to its upper surface. To the east, the main body of the flow consists of massive, locally flow-banded rhyolite. Trace the base of the rhyolite northeastwards, back to the small stream, noting how the basal rhyolite blocks ploughed into the unlithified top of the underlying tuffs, which formed the sea floor over which the rhyolite flowed.

From (12) it is only a short ascent to the ridge of Bwlch Cyfryw drum, which extends between Carnedd Dafydd and Carnedd Llewelyn and affords some of the finest views in Snowdonia. From the ridge a direct descent may be made down Nant Fach over the Llafar moraine, to cross the Afon Llafar and rejoin the footpath from Ysgolion Duon to Gerlan at (2). Keen walkers may choose to climb to the summits of Carnedd Llewelyn and Foel Grach, examine the castellated outcrops of basalt between the peaks and then return to Gerlan via the valleys of the Afon Wen and Afon Caseg. This route allows examination of the stone polygons (Fig. 10) on the col south of Foel Grach. An alternative route back to Gerlan lies over the summit of Carnedd Dafydd, descending the north-western ridge to the lower part of the Llafar Valley, near (1); or for parties that can arrange to be picked up on the A5, Ogwen Cottage can be reached by following the path over Carnedd Dafydd and Pen yr Ole Wen.

4 Nant Ffrancon

Duration: ½ day.

Nant Ffrancon, along which the A5 trunk road runs, has long provided one of the main routes for travellers crossing the Snowdon massif. The spectacular scenery along the valley is the result of glacial erosion; and the variety of glacial landforms, together with easy access, make a short excursion to the upper part of the valley particularly attractive. It can be conveniently linked to Excursion 5, Cwm Idwal.

This excursion (Figs 18, 19) begins at the conveniently situated Cwm Idwal car park [648 603] (see Excursion 5). Roadside parking for coaches is available by Llyn Ogwen on the main A5 trunk road.

The classic U-shaped cross-section of Nant Ffrancon is at its most impressive when viewed from the head of Rhaeadr Ogwen [648 605] (**1**), 200 m north of the car park. At Rhaeadr Ogwen the stream flowing from Llyn Ogwen cascades over a large, glacially sculptured rock step to the flat valley floor of Nant Ffrancon. The contrasting landforms of the steep east and west facing slopes of the valley should be noted. The west-facing side of the valley is topped by crags above the steeply inclined rockfalls and screes over which the route of the A5 passes. Along the east-facing side, however, seven separate cwms have been excavated into the valley side, high above the valley floor (Figs 18, 19). During the last stages of the glaciation of Nant Ffrancon, about 10 000 years ago, small glaciers occupied these cwms, and their formation only on the east-facing slopes of the main valley was controlled by meteorological factors. The prevailing winds in Snowdonia during the glacial period are believed to have been from the southwest, as they are today, and snow accumulated preferentially in the lee of hills, in hollows on northeast-facing slopes, which were exposed to little direct sunlight. As a result, glaciers formed in the hollows, which were then enlarged by subsequent glacial erosion into cwms. A glance at the 1:50 000 topographic map of Snowdonia shows that most of the cwms throughout the area face between the north and east.

To reach (**2**), walk or drive down the minor road on the western side of Nant Ffrancon. This road, which is unsuitable for coaches, was originally constructed in 1791 to link the Penrhyn slate quarry with the quarry-owner's estates at Capel Curig. It was later modified by Thomas Telford as part of his famous London–Holyhead route. The locality [638 623] is situated on a large alluvial cone, deposited by the stream descending from Cwm Bual. This stream cuts through the moraine at the northern side of the cwm, and the cone is built of material from the moraine together

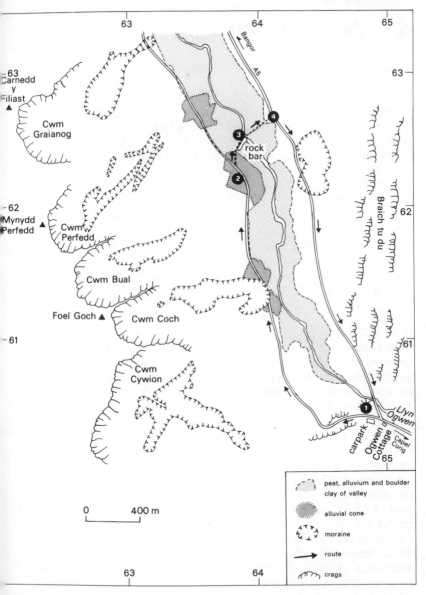

Figure 18 Sketch map. Excursion 4, Nant Ffrancon.

Figure 19 Annotated photograph of part of route, Nant Ffrancon (Excursion 4). Locality (2) is situated on an alluvial cone (IGS photograph).

with mudstone fragments derived from the adjacent slopes. The cone originally extended into a delta in one of the two ribbon lakes that formerly occupied the floor of the valley. The lakes were situated behind rock bars, one of which crossed the valley opposite Ty Gwyn Farm.

Part of this rock bar is exposed and can be examined along a footpath leading from the road at the cone to a footbridge across the Ogwen River [639 625] **(3)**. The rock (intrusive porphyrite) was more resistant to ice erosion than the surrounding sedimentary rocks and hence remained as an upstanding feature. Good glacial striations on the surface of the rock bar can be examined on an outcrop in the stream just below the footbridge. Valley infill deposits – including glacial till, lake alluvium and an older vegetation layer represented by peat – are exposed in the river bank on the upstream side of the footbridge.

From the footbridge there is a fine view upstream to the rock step at the head of Nant Ffrancon and beyond to the Glyders. Downstream, a second rock bar, carved into well-developed roches moutonnées, can be seen below the conspicuous tips of the Penrhyn slate quarry. This rock obstruction provided a convenient route for Telford's old road to cross to the eastern side of the valley, where it now joins the more recently constructed A5.

Continue on the footpath from the footbridge to the A5 [642 627] **(4)**. Looking back at the western side of the valley, the main features of the high cwms on the east-facing slopes can be examined. Each has mounds of hummocky moraine extending across the open side and downslope from the cwm. In the case of Cwm Coch (Fig. 18) the downslope extension of the moraine has been deeply dissected by a stream. The western and northern walls of Cwm Graianog (Fig. 19) expose a ripple-marked bedding surface within the Carnedd y Filiast Grit; the ripple marks are clearly visible, even though the cwm is 1 ½ km from the viewpoint.

The footpath along the side of the A5 leads back to the head of Nant Ffrancon and to the Cwm Idwal car park.

5 Cwm Idwal

Duration: ½ day.
Geological map: IGS 1:25 000 Geological Special Sheet, Central Snowdonia.

The rocks that occur in Cwm Idwal extend from the Pitts Head Tuff through to the Bedded Pyroclastic Formation. They include a wide range of volcanic rocks, dominated by acidic ash-flow tuffs and both acid and basic lavas. In parts of the sequence the intervening sedimentary rocks are richly fossiliferous and locally show distinct enrichment with volcanic debris. The rocks have been folded into a symmetrical syncline, a cross-section of which is most clearly displayed in the back wall of the cwm, in the highest beds of the sequence. The area graphically displays the erosional and depositional features of glaciation, and in addition it is a classic locality for rare Alpine flora. No plants should be picked or uprooted, since many of the species are so rare that they are protected by law. The area of the cwm is a nature reserve, where the use of geological hammers is prohibited – and, it should be said, is unnecessary. Care should be taken by visitors not to leave any indication of their visit – especially, in the clearly vulnerable areas, not to widen or damage the footpaths.

The excursion (Figs 20, 21, 22) begins at Idwal Cottage [648 603], where a convenient car park lies adjacent to the Youth Hostel just off the A5. Coaches may be parked in a lay-by at the side of the A5 opposite Llyn Ogwen.

The Pitts Head Tuff is clearly displayed in the road cutting (1) on the A5 150 m northeast of the car park, although, because of its situation, it is not convenient for examination by a large party. Here it comprises a lower thick ash-flow tuff member and an overlying thin ash-flow tuff. The outcrop is on the western limb of the Idwal Syncline, and the beds dip steeply to the east. The base of the tuff is well exposed towards the northern end of the cut, where it rests on sandstones, and the basal 2 m is rich in small feldspar crystals. The central part of the ash-flow tuff is massive and well jointed. These joints are probably cooling joints, which developed normal to the base of the flow following its emplacement, although now, as a result of the folding, they are at an acute angle to the base. The tuff is grey–green and highly silicified. On fresh surfaces it is possible to discern a welding foliation accentuated by the alignment of dark

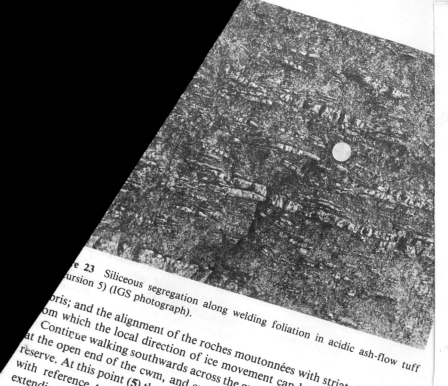

Figure 23 Siliceous segregation along welding foliation in acidic ash-flow tuff (Excursion 5) (IGS photograph).

...oris; and the alignment of the roches moutonnées with striated surfaces, from which the local direction of ice movement can be deduced.

Continue walking southwards across the grassy mounds of morainic drift at the open end of the cwm, and cross the stile into the area of the nature reserve. At this point (**5**) the solid geology of the cwm should be considered with reference to the published geological map of the area. The view extending from the east to the southwest indicates very clearly the sequence from the Capel Curig Volcanic Formation on Tryfan to the Pitts Head Tuff at the back of Cwm Cneifon (the Nameless Cwm), the Lower Rhyolitic Tuff Formation in the massive crags above the Idwal Slabs (Fig. 21) and the basalts in the core of the Idwal Syncline in the back wall of the cwm about Twll du (Fig. 22).

Also from this locality, the cirque morphology of Cwm Idwal, with its varied range of drift deposits, can be seen. Note particularly the classic ...oundy moraines at the lakeside towards the innermost end of Llyn Idwal, ... huge block fall from the crags about Twll du at the back of the cwm, ... the screes and alluvial cones. Particularly well featured are the small ...ues of Cwm Cneifon and Cwm Clyd high above, and on either side of, ... Idwal.

...llow the track along the eastern margin of the lake. The steep west-

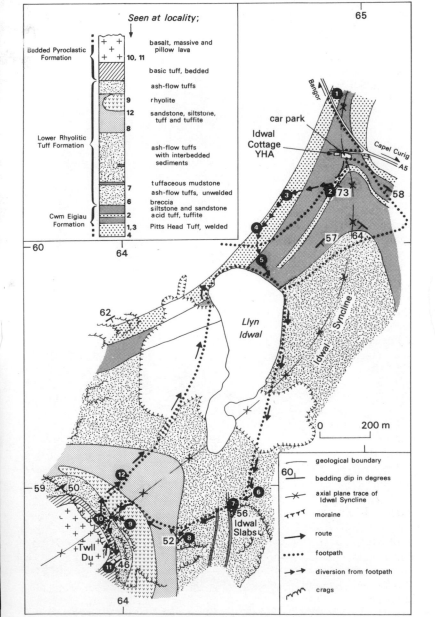

Figure 20 Sketch map. Excursion 5, Cwm Idwal.

Figure 21 Annotated photograph showing route, Idwal Slabs (Excursion 5).

Figure 22 Annotated photograph showing route, Twll du (Excursion 5). LR, Lower Rhyolitic Tuff Formation; R, rhyolite; P, Bedded Pyroclastic Formation; B, basalt (IGS photograph).

fiamme (Fig. 4). Th
cates movement wit
ponents were still
dstones showing good
ces of large fossil brach
Return to the car park and
rough the old quarry (2). Th
nd cross-bedded, fine-grained,
hich is a mixture of volcanic
mudstone; the thin alternating
association is common in the seque
across the whole of Snowdonia; it rep
distant vent, which was carried in the pr
area into the water in which the muddy se
horizon occurs between the Pitts Head Tuff
Formation. Other associated sediments, seen
quarry, are predominantly coarse sandstones i
volcanic ash that settled through water. The
siliceous character of the tuffs made them suitable
sharpening metal implements, and this was the reaso

Take the path from the quarry across a smal
westwards 200 m to the distinctive ice-scoured crags (3)
tuff of the Pitts Head Tuff is well exposed, both overlain b
sandstones. Similar features to those of locality 1 occur, b
surfaces it can be seen that the welding foliation has been a
silicification along the closely spaced foliation planes (Fig. 2
centre of the section this foliation defines a fold, in which
prominent zone of spherical siliceous nodules, about the size of ta
balls, near the base. Such nodules are often found in the acidic r
Snowdonia and may have been produced in several ways. The
common explanations, depending upon variety, are (a) that they repres
infilled bubbles of gas in the tuffs or lavas and (b) that silica (quart
aggregated about a nucleus and grew over the original tuff fabric.

Walk 200 m southwestwards along the ridge of low crags to the
conspicuous outcrop of (4), noting on the way the clear evidence of ice
erosion. This is particularly apparent about (4) at one of the most
photographed roches moutonnées in Snowdonia (Fig. 9). It is prominen
enough to provide a view southwestwards into Cwm Idwal and northwes
wards down the Nant Ffrancon valley to Bethesda and further to the Me
Straits and Anglesey. At this vantage point it is well worth spending so
time considering the glacial geomorphology of the area, including
valley profile of Nant Ffrancon; the line of northeast-facing cwms hig
the southwestern side of the valley; the nearby spreads of moundy mo

facing slope above the track is eroded in the basal tuffs of the Lower Rhyolitic Tuff Formation, which form a prominent dip and scarp feature in the Idwal Slabs at the far end of the track (Fig. 21). The basal part of this formation, seen at (6) just before the slabs are reached, is characterised by breccias, comprising blocks of acid tuff, vesicular basalt and sandstone in a matrix of finer volcanic debris. Traversing to the scarp, at the edge of the Idwal Slabs, the breccia can be seen grading upwards through acidic ash-flow tuffs, with zones of breccia, into massive ash-flow tuffs on the lowest of the slabs, which form a dip slope (46°) to the west.

Continuing southwestwards along the footpath below the Idwal Slabs to (7) (Fig. 21), it can be seen that higher in the sequence the Lower Rhyolitic Tuff Formation consists of massive bedded ash-flow tuffs, tuffs and interbedded sediments rich in volcanogenic debris. A fine-grained tuffaceous mudstone, 1·5 m thick, forms a distinctive weathered-out feature extending up the crags near the centre of the section. Above the mudstone the influence of the sedimentary environment increases sharply, with many of the tuffs showing cross-lamination, graded bedding and large carbonate nodules up to 1 m in largest dimension. It is thought that the ash-flow tuffs in the lower part of the Lower Rhyolitic Tuff Formation were erupted in the south near Snowdon and possibly flowed from the land into the sea. The ash-flow tuffs are unwelded and shard rich, with few fragmented feldspar crystals. Also in the sequence above the Idwal Slabs, bands of fine-grained tuff occur. These are interpreted as air-fall water settled fine ash that was probably transported in the prevailing winds from the same vent that produced the ash flows.

The top of this lower unit of the Lower Rhyolitic Tuff Formation is well featured in outcrops just above the footpath to the south of the Idwal Slabs (8), where it is overlain by siltstones and sandstones with thin tuff and tuffite bands. (These will be examined at (12).)

Follow the track, across the steep scree-covered slope, towards Twll du at the back wall of the cwm in the core of the syncline (Fig. 22). The steep crags (9) are in dark blue–grey rhyolite, with distinctive columnar joints developed in places. The rock has a flinty character; and in outcrop, especially on weathered surfaces, fine contorted flow-banding and brecciation resulting from flow can be examined. Also, with the discriminate use of a hammer on loose blocks only, and not on *in situ* rock, it is possible to obtain specimens showing fine interfering arcuate fractures (**perlitic** fractures, Fig. 2); these are a characteristic cooling contraction feature of glassy acidic rocks, such as obsidian, and are clearly visible under a hand lens.

Follow the footpath to just below the cleft of Twll du (10) where the rhyolite is overlain by massive bedded acidic tuffs composed of shards, feldspar crystals and, in places, much sedimentary material. These tuffs are interpreted as the deposits of subaqueous flows developed from the downslope slumping of earlier-deposited ash and sediment. At this locality the junction between these rocks at the top of the Lower Rhyolitic Tuff Formation and the basic tuffs at the base of the Bedded Pyroclastic Formation is well exposed. The basic tuffs and tuffites are distinctively green, fine to coarse grained and well bedded, and in places they show cross-laminations and well-developed ripple marks. The alternation of beds of coarse and fine debris reflects changes in the nature of the pyroclastic materials erupted from relatively local vents. In the steep cliffs on either side of Twll du and above the path that extends up to Glyder Fawr a clearly defined junction between these tuffs and the pillowed basalts in the core of the syncline can be seen (Fig. 22). On these crags, referred to in climbers' handbooks as the Hanging Gardens, a rich Alpine flora is established; this is particularly impressive in early summer days, the flora taking advantage of the calcareous character of the altered basic rocks. On no account should the plants be disturbed. Many blocks of the pillowed basalt occur in the rockfall below the crags, where they are most easily examined.

Continue on the Glyders footpath to (**11**), where a particularly impressive ripple-marked bedding plane in the basic tuffs is exposed; this suggests that the tuffs were deposited in fairly shallow water. The tuffs here include

Figure 24 Convolute lamination in bedded tuffite and siltstone (**12**, Excursion 5) (IGS photograph).

some fine air-fall acid tuff bands, representing contemporaneous acid volcanicity probably some distance away (see Excursion 9).

Return down the path to the base of Twll du (9), where two paths diverge. Follow the left hand track down through the blockfall. On a steep section of the track, where it emerges from the blockfall and crosses solid rock at (12), bedded siltstones with fine to coarse-grained tuffites of the upper part of the Lower Rhyolitic Tuff Formation are well exposed (Fig. 24). Some layers, bounded by even bedding planes, are internally contorted as a result of either slumping or **de-watering**. De-watering is the rapid migration and loss of water held within the unconsolidated sediment or ash, due to a sudden change of confined pressure. A probable cause would be shock waves associated with earthquakes, which must have been commonplace in this volcanically active area.

From (12), as from all the localities higher in the slope, the elongate form of the classic moraines on the western margin of Llyn Idwal are well displayed. These moraines are renowned, for they were part of the evidence that convinced Charles Darwin of the previous existence of glaciation in North Wales and hence elsewhere in the British Isles. Actually, Darwin failed to notice this evidence on his first visit to Cwm Idwal in 1831; but a decade later, following a second visit, he wrote: 'these phenomena are so conspicuous that a house burnt down did not tell its fate more plainly than did this valley'. The moraines extend well into the back of the cwm, and on their inner margins they are overlain by large alluvial cones extending down from the streams off the high crags.

Follow the track over the moraines, keeping carefully to its central line to avoid unnecessary erosion of these quite remarkable features, and return to Idwal Cottage.

6 Ogwen Cottage to Cwm Tryfan

Duration: 1 day.
Geological map: IGS 1:25 000 Geological Special Sheet, Central Snowdonia.

This excursion is to examine the sequence of sedimentary and volcanic rocks in the Cwm Eigiau Formation and part of the Llewelyn Volcanic Group, below the Lower Rhyolitic Tuff Formation, in the well-exposed section extending from the Gribin ridge to Cwm Tryfan. These rocks form the common limb between the northward-trending Idwal Syncline to the west and the Tryfan Anticline to the east – the two major fold structures of the area. The terrain has been dramatically scoured by the Pleistocene ice, and on route many examples of differential erosion in hard and soft strata are seen and impressive moraines traversed. A good hill walk of about 8 km is involved.

The route (Fig. 25) starts at the car park adjacent to Idwal Cottage Youth Hostel [648 603].

From the car park, follow the Cwm Idwal footpath at the back of Ogwen Cottage. Immediately behind the cottage, crags in the Pitts Head Tuff, on the east limb of the Idwal Syncline, lie east of the path. On the northern side of the crags the base of the acidic welded ash-flow tuff is well exposed. It is generally concordant with the underlying sandstones, although local irregularities occur where rafts of pebbly sandstone are caught up in the tuff.

About 100 m southeast of the cottage (1) a prominent feature east of the footpath provides a good section through the Pitts Head Tuff. The tuff is characteristically welded, with good eutaxitic foliation and fiamme. Near the centre of this section a raft of sandstone, locally conglomeratic, breaks the tuff sequence. At the top the tuff grades upwards into sandy tuffite – the reworked top of the tuff. These sandstones, well exposed to the southeast, show very well-defined cross-bedding and in places are rich in brachiopods – features indicating shallow water deposition.

Continue along the footpath, passing the branch leading southwestwards to Cwm Idwal. On reaching the large lateral moraine about 500 m southeast of (1), traverse southwestwards to the crags (2) [652 597] in the Lower Rhyolitic Tuff Formation at the northern end of the Gribin

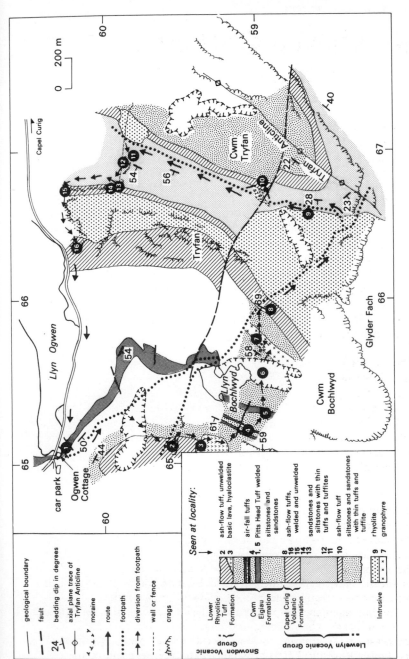

Figure 25 Sketch map. Excursion 6, Ogwen Cottage to Cwm Tryfan.

Map legend:

| geological boundary |
| fault |
| 24 bedding dip in degrees |
| axial plane trace of Tryfan Anticline |
| moraine |
| route |
| footpath |
| diversion from footpath |
| wall or fence |
| crags |

Seen at locality:

2	ash-flow tuff, unwelded basic lava, hyaloclastite
3	air-fall tuffs
1, 5	Pitts Head Tuff welded siltstones and sandstones
4	
8	ash-flow tuffs, welded and unwelded
16	
15	
14	
13	sandstones and siltstones with thin tuffs and tuffites
12	
11	
10	ash-flow tuff siltstones and sandstones with thin tuffs and tuffite
9	rhyolite
7	granophyre

Snowdon Volcanic Group:
- Lower Rhyolitic Tuff Formation
- Cwm Eigiau Formation
- Capel Curig Volcanic Formation

Llewelyn Volcanic Group

Intrusive

Map labels: Capel Curig, Llyn Ogwen, car park, Ogwen Cottage, Llyn Bochlwyd, Tryfan, Cwm Tryfan, Tryfan Anticline, Cwm Bochlwyd, Glyder Fach

Scale: 0 — 200 m

ridge. The base is well featured, with the underlying siltstone, interbedded with thin cross-bedded sandstone, occupying the negative feature extending up the steep slope. At its base the tuff is pale grey and rich in feldspar crystals and includes small clasts drawn out along the cleavage. Above, prominent zones of tuff breccia – with clasts of acid tuff, basalt and sandstone, up to 1 m across – can be traced up on to the top of the crags. The main body of tuff, occurring in the slope extending down into Cwm Idwal, comprises massive medium-grained acid tuff with a shard-rich matrix, few feldspar crystals and small clasts. The tuff displays a prominent set of vertical cooling joints. This sequence, in the lower part of the Lower Rhyolitic Tuff Formation, can be followed to the south to the Idwal Slabs (Excursion 5).

Trace the base of the tuff upslope for about 150 m, to a point where the tuff overlies a basaltic pillow breccia (**3**). The base of the breccia overlies siltstones and sandstones and can be traced easily along the ridge; its top corresponds with the topographic slope down towards Llyn Idwal. Good sections along joint surfaces, showing whole and fragmented pillows set in a matrix of fragmented chloritised basaltic glass, should be examined. The basaltic breccia extends southwards across the lip of Cwm Cneifon (the Nameless Cwm) on to Glyder Fach. Its thickness is variable, and it

Figure 26   View eastwards from the Gribin ridge, across Llyn Bochlwyd to the west-facing flank of Tryfan (Excursion 6) (IGS photograph).

represents local accumulations of basalt prior to the eruption of the extrusive acidic ash-flows of the Lower Rhyolitic Tuff Formation.

This point on the Gribin ridge is a splendid position from which to view the area; Cwm Bochlwyd and Tryfan lie to the east (Fig. 26), Cwm Idwal and Nant Ffrancon to the west, and Pen yr Ole Wen and the Carneddau to the north. The opportunity should be taken to study the major geological structures and, of course, the massive glacial features.

Cross the depression on to the scarp (4) overlooking Cwm Bochlwyd. The scarp consists of a well-bedded sequence of sandstones, tuffs, tuffites and interbedded siltstones. The acidic tuffs are well-bedded, air fall and locally reworked. The sandstones are cross-bedded and in places include much volcanic debris. The alternating sedimentary rocks probably reflect changes in sediment supply related to the instability associated with the volcanicity, which is represented by the fine-grained air-fall ash that settled into the depositional environment.

Descend to the small peat flat (5) bounded on its southern side by a scarp exposing the Pitts Head Tuff and the underlying and overlying sandstones. The tuff is clearly welded, with fiamme extended in the foliation. The associated sandstones are massive and flaggy bedded, with locally clear evidence of soft sediment deformation of the internal laminations. The upstanding ridge feature formed by the tuff and sandstone extends southwards on to the Gribin ridge.

Towards the east the route traverses the moraine (6) which is exceptionally well featured at its margin; its extent is clearly defined by the irregular mounds of very large blocks.

At (7) a prominent ice-scoured crag exposes massive and flaggy cross-bedded sandstone intruded by a thick well-jointed granophyre sill. It is instructive to trace the boundaries of the sill through the outcrop on the southern side of the path extending to Bwlch Tryfan. From this locality the view to the north along the western flank of Tryfan demonstrates very clearly the differential erosional effects on hard and soft strata. The sandstones and volcanic rocks form the scoured crags, and the siltstones and mudstones form the grassy slopes and intervening depressions.

Follow the path towards Bwlch Tryfan. At (8) the top of the Capel Curig Volcanic Formation, comprising bedded clast-rich acid tuff with accretionary lapilli, is exposed dipping steeply to the northwest. The contact strikes along the base of the steep face of Glyder Fach to the south and along the prominent features at the top of the high grassy slope on Tryfan to the north.

Continue southeastwards along the track at the base of the rhyolite crags and scree-covered slopes of Glyder Fach. Below the path, screes abut on hummocky lateral moraines that extend to the northeast down to lower ground. At the scree, before the lip of Cwm Tryfan is reached, turn northwards downslope to follow the main path down the cwm. The crags to the west are mainly of intrusive rhyolite, although the basal contact with sandstones and siltstones is exposed. These sediments are below the Capel Curig Volcanic Formation.

Just south of the minor stream from the southwest (9) examine the well-developed folded flow bands in the rhyolite, with both tight and open folds that developed during flow. From the vantage point east of the path the closure of the Tryfan Anticline can be seen in the cwm. On the gently dipping eastern limb there is a thick sequence of siltstones, with thin tuff and tuffite beds, passing up to sandstones, also with thin beds of tuff and tuffite. These same rocks extend around the southern closure of the fold and dip westwards in the outcrops about the vantage point on the western limb.

East of where the path crosses some small moraines an unwelded acid tuff, up to 50 m thick, forms a prominent feature (10). This distinctive tuff is massive at its base and bedded at the top, with small garnet crystals dispersed throughout.

Northwards, the path follows the strike of the beds, and immediately to the west a prominent sandstone crops out, with thinly bedded, fine-grained, air-fall tuff and tuffite at its base. Alternating sandstones and siltstones form the well-featured lower slopes of Tryfan. The imposing high crags of the mountain comprise welded tuffs of the Capel Curig Volcanic Formation and an intrusive rhyolite sill – the continuation of the intrusion seen to the south. The sill intrudes the sediment split between the lowest two tuff members of the formation on Tryfan. Sediments at the base of the rhyolite form the well known Heather Terrace across the face of the mountain. From below, the rhyolite is clearly recognised by its columnar jointing. Eastwards, looking into the cwm, the well-bedded rocks forming its eastern wall are sandstones underlying the garnetiferous tuff. There is a marked facies change across the structure, for on the western limb siltstones predominate.

Proceed down the path, noting the ice-polished bedding surfaces in the bottom of the cwm and, to the north, the prominent terminal moraines. Near the fenced lower ground, the poorly featured area east of the path is an acidic intrusion, largely of rhyolite although with patches of microgranite.

Bear westwards along the southern side of the fence, which here follows the course of a cleaved dolerite dyke, to examine the dip section of some 300 m of sediments underlying the Capel Curig Volcanic Formation. These are predominantly sandstones, with siltstones and tuffites. The sandstones comprise numerous flaggy and massive beds with cross-bedding. At (11) small-scale trough cross-bedding, accentuated by heavy mineral laminae, is well displayed. Both trough and planar tabular cross-bedding can be determined at **(12)**, and westwards the latter type is common. Sandstone beds and thicker composite units are generally separated by siltstone, and there is little evidence of erosive downcutting bases. They are interpreted as fluvial sandstones deposited in shallow water. It can be confirmed along the section that the prevailing transporting current direction was from the northwest. At **(13)** the sandstones are moderately inclined; but in the direction of dip the attitude varies, the beds are subvertical, and individual beds show discontinuity.

The base of the lowest tuff member of the Capel Curig Volcanic Formation, locally parallel to the steeply inclined bedding, can be examined at **(14)**. It extends up the crags oblique to the regional strike, and the tuff locally lobes down into the sandstones or is intruded by flames of the underlying sandstone. The disturbed bedding of the underlying sandstone results from the stresses applied by rapid emplacement of tuff on to wet sediment. At **(14)** the base of the tuff is grey–green and cleaved and, above, the massive tuff is packed with siliceous nodules. These are ellipsoidal, flattened in the plane of cleavage during tectonism.

This tuff crops out downslope **(15)**. Here the welding foliation can be discerned on close inspection, and wispy bedding is apparent in the top metre. The overlying beds of tuffite and dark grey siltstone have been preferentially eroded to form a negative feature. The overlying intrusive rhyolite forms much of the northern face of the mountain. The flow banding near the base and top is nearly parallel to the contacts, whereas in the central part of the intrusion it has a variable attitude with prominent folding. Towards the top of the sill silica recrystallisation has produced thick coalesced banding and, locally, nodule development.

The overlying acid tuff **(16)** succeeds the tuff of **(15)**, but here is separated from it by some 300 m by the rhyolite. It is unlike the lowest member, being well bedded, coarse grained, poorly sorted and unwelded. Cross-bedding can be examined on crags about the locality. Whereas the lowest member of the formation is a single large ash-flow from a distant source, probably the north, the bedded tuff is locally derived from a source vent in the Glyders area, where the tuffs are agglomeratic.

Return to Ogwen Cottage by the main road, situated just below **(15)**.

7 Gallt yr Ogof and Llynnau Mymbyr

Duration: 1 day.
Geological maps: IGS 1:25 000 Capel Curig and Betws y Coed (SH 75); and
IGS 1:25 000 Geological Special Sheet, Central Snowdonia.

This excursion is to examine the Capel Curig Volcanic Formation and associated sediments in two areas: Gallt yr Ogof, in the Ogwen Valley; and the northern side of Llynnau Mymbyr, Capel Curig. The excursion is in two parts, each of which may be followed independently, but the interpretation of the sequences that follows the itineraries includes detail from both localities.

(a) Gallt yr Ogof

Gallt yr Ogof is the ridge extending southwards from the A5 on the eastern limb of the Tryfan Anticline. The excursion is to examine the crags at the northern end of this ridge [693 596]. These can be reached by the 'old road' from Gwern Gof Isaf Farm [685 601], where car-parking facilities are available (Fig. 27).

To reach the first locality, turn south of the 'old road' at its junction with the public footpath from the A5.

The crags (**1**) comprise fine to coarse-grained sandstones underlying the Capel Curig Volcanic Formation. The sandstones occur in thin flaggy beds and composite units, up to 4 m thick, with fine green–grey siltstone intercalations. Some of the thin sandstone beds, low in the section, are graded; thicker beds show large-scale tabular cross-bedding; and the composite units contain numerous cross-bedded sets. These sandstones are interpreted as fluvial deposits in a shallow water environment, where the background sedimentation is represented by the finer siltstone partings.

Around the main crags, higher in the slope, a prominent face (**2**) allows detailed examination of the sedimentary structures. Sandstone beds, with well defined cross-bedding and fine and pebbly alternations, are separated by evenly laminated tuffitic siltstone, up to 10 cm thick. Small load casts of the overlying sandstone intrude the siltstone, and careful examination reveals small V-like casts of sandstone at the top of the siltstone. These are infilled **desiccation cracks** (Fig. 5), indicating that the unconsolidated silt was temporarily free of its water cover.

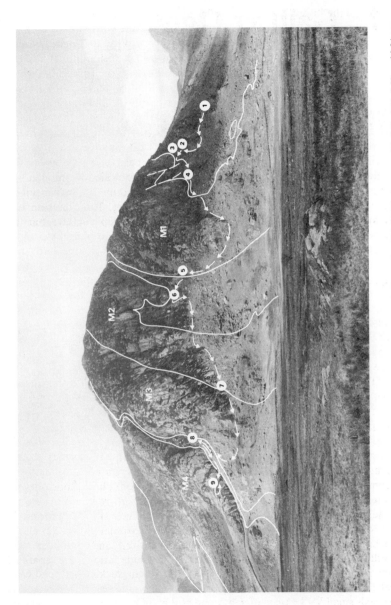

Figure 27 Annotated photograph showing route, Gallt yr Ogof (Excursion 7a) (IGS photograph). M1, M2, M3 and M4 are the 1st, 2nd, 3rd and 4th members, respectively, of the Capel Curig Volcanic Formation.

At (3) the lowest ash-flow member of the Capel Curig Volcanic Formation overlies the sediments conformably. However, when traced both up the gully and eastwards along the base of the crags the tuff sharply transgresses the underlying sediments. This downward intrusion of tuff into the sediments is interpreted as a large load cast. Locally, the contact is slightly displaced by faulting, but at (4) the basal irregularities can be clearly traced. Here the sandstone near the contact is massive, with ill-defined bedding, and includes isolated pods of tuff. Further to the east the adjacent sandstone is totally disturbed and intrudes the tuff. Along the crags, in the direction of dip, textural variations in the tuff can be observed. It is fine grained, green–grey at the base but becoming paler upwards, where the welding foliation is apparent. Isolated siliceous nodules occur throughout, and locally the nodules are closely packed in zones (Fig. 28). As a result of the highly irregular base the thickness of the tuff is variable, but approximately 130 m occur between the basal contact (4) and the top of the flow (5). In contrast to the base, the top is planar, overlain by 5 m of reworked unwelded tuffs rich in feldspar crystals, which pass upwards into silty tuffite and then fossiliferous grey siltstones.

Here the overlying second tuff unit of the formation comprises two ash-flow tuffs. The base of the lowest tuff lobes down into the underlying

Figure 28   Siliceous nodules at the top of an acidic ash-flow tuff (Excursion 7a) (IGS photograph).

siltstones. These lobes extend downwards, to 50 m in places, and part of such a lobe can be traced about (6). The base of the parent tuff can be seen 40 m to the east. The siltstone at the contact is indurated and crowded with feldspar crystals incorporated into the silt at the time of tuff emplacement. The tuff, 60 m thick, has a cleaved crystal-rich base, a massive central part with a well-defined eutaxitic welding fabric and zones of siliceous nodules, and an upper fine-grained part. This tuff is directly overlain by the third ash-flow tuff of the formation, and the contact is marked by a prominent feature at (7).

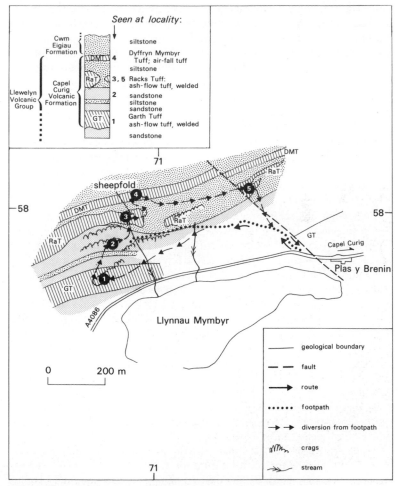

Figure 29   Sketch map. Excursion 7b, Llynnau Mymbyr.

The lower few metres of this third tuff comprises alternating bands of coarse and fine tuff, which pass up into massive tuff with coarse parts containing abundant rock fragments and crystals. The upper part of the tuff forms several well-defined bedding features in the grassy slopes below the crags. The top is exposed in the gully (8), overlain by 5 m of fossiliferous siltstone.

All three ash-flow tuff members to this point have similar overall characters. However, the succeeding member (9) comprises tuffs that differ and reflect a change of source. The lowest 20 m are an unwelded ash-flow tuff. The basal zone is cleaved and rich in feldspar crystals. Above, the tuff is massive, with wispy bedding lamination in the upper half. Cavities due to the weathering of calcareous concretions are common. Above this flow there are some 15 m of coarse and fine-bedded tuffs showing cross-bedding and downcutting channels; the top 5 m are rich in accretionary lapilli.

(b) Northern side of Llynnau Mymbyr

This is the type area of the formation. It was originally described by Howel Williams, who considered the acidic ash-flows to be rhyolites emplaced in a submarine environment. The route (Fig. 29) is easy; parking is possible in the large layby overlooking the lake. The route begins at the marked Glyders footpath [714 578] near Plas y Brenin.

Follow the Glyders footpath westwards, bearing southwards to the crags (1) in the Garth Tuff − the lowest member of the Capel Curig Volcanic Formation. Here the base of the tuff is not exposed. The lowest exposed tuff is massive and cream−white, with a few sodic feldspar crystals, and on fresh surfaces a strong welding fabric can be determined with a hand lens. Isolated siliceous nodules occur near the base of the crags. Higher in the crags a crude bedding foliation is apparent on the weathered surface of the massive tuff, and upwards a gradation into reworked flaggy cross-bedded tuffs can be traced. These reworked tuffs include shards and thin-walled pumice fragments that are evidently unwelded. The top of the tuff is marked by a clearly defined feature accentuated by glacial scour. The tuff is overlain by cross-bedded sandstones with interbedded fine sandy siltstone.

Higher in the slope (2) the sandstones have been quarried, presumably for building stone. Here, thin clast-rich bands and layers containing small brachiopod shells can be examined.

Walk eastwards to the small stream and climb towards the small sheepfold (3) at the top of the feature formed by the Racks Tuff. The tuff forms the massive crags on the western bank of the stream, and here it is clearly a single welded ash-flow unit, defined by the distinctive vertical cooling joints. Near the sheepfold the upper part of the tuff is locally crowded with siliceous nodules.

Trace the tuff eastwards, along the feature, and notice that within 150 m it has wedged out and the mudstones, above and below, form an uninterrupted sequence. From this point, crags formed of isolated bodies of the tuff can be seen along strike further east. The feature at the top of the Racks Tuff also forms a good vantage point from which to view the ridges of the Lower Crafnant Volcanic Formation (Excursion 9) to the east, the feature formed by the dolerite at the top of Moel Siabod and the anticline in the crags about Llynnau Mymbyr to the south, and the Snowdon massif to the west.

Cross the negative feature, in drift-covered mudstones and siltstones, to (4). Here, small exposures occur in bedded tuffs and tuffites of the Dyffryn Mymbyr Tuff – the highest member of the formation. These are characterised by the inclusion of both whole and fragmented accretionary lapilli, which indicate that the volcanic activity had a subaerial expression.

From this point the route crosses the scoured feature at the top of the Racks Tuff, eastwards along strike to an isolated body of Racks Tuff (5). Here it is recommended that time be spent examining the contact exposed on the northeastern margin. The contact is extremely irregular; fingers of the tuff extend into the adjacent mudstones and siltstones, which in places include feldspar crystals. The mass of the tuff is demonstrably welded, and thin sections have shown that the tuff is welded up to the contact. In the vicinity of the contact the bedding in the sediments is much disturbed.

Return to the footpath and proceed back to the main road near Plas y Brenin.

Interpretation

The lower two tuff units at Gallt yr Ogof and Llynnau Mymbyr are characterised by irregular bases, welding down to the lower contact, siliceous nodules and reworked tops of unwelded shardic debris. The irregular bases are interpreted as having developed from the rapid emplacement of the ash-flow tuff on water-saturated sediment. The sediment yielded, probably as a consequence of the shock of the rapid emplacement of the thick ash-flows, but possibly also in response to unequal loading and seismic shock associated with the volcanicity; and the tuff collapsed downwards, with concomitant upward intrusion of remobilised sediment into the tuff. The characters of these bases vary, from the regular upward incursion of sediment on Gallt yr Ogof to the highly irregular downward lobing at Llynnau Mymbyr and to the south.

As a result of a detailed study (Howells *et al.* 1979) these lower two tuff units are interpreted as having been erupted on land to the north of the Ogwen Valley. From there they extended through a transitional zone in the Gallt yr Ogof area into a marine environment at Capel Curig and to the south.

The upper tuff unit at Capel Curig and Gallt yr Ogof is petrographically and lithologically distinct; it is considered to represent an accumulation of primary and secondary-emplaced ash-flows, whose source lay in the Glyders area. The frequent occurrence of accretionary lapilli in these tuffs indicates that this volcanicity had, at least in part, a subaerial expression.

8 Llanberis Pass

Duration: ½ day.
Geological maps: IGS 1:25 000 Geological Special Sheet, Central Snowdonia.

This excursion is mainly to examine the Pitts Head Tuff and the overlying strata to the base of the Lower Rhyolitic Tuff Formation, near Pont y Gromlech in the Llanberis Pass. It is centred on the most popular climbing crags in the pass. The excursion route is short and fairly easy, although it involves some traverses across block-strewn slopes and small screes. It provides a strong feeling of the dramatic splendour of central Snowdonia; apart from the view through the northwestern end of the pass, the eye is drawn upwards to the high crags and ridges.

The excursion route (Figs 30, 31) begins at Pont y Gromlech [630 566], where the Pitts Head Tuff forms the large crags on the southern side of the road. Parking, although restricted, is available in the roadside laybys in the Llanberis Pass.

Cross the stile on to the small track that extends along the western edge of the crags (1). Here the acidic ash-flow tuff is intruded by a thin basaltic dyke. The tuff is white weathered, with siliceous segregations accentuating the welded foliation and chloritic fiamme prominent on the weathered surface. The basalt dyke cuts across the foliation and can be traced to the small crags to the west.

Traverse along the edge of the main crags (2), noting the strongly welded character of the ash-flow tuff and a prominent zone of siliceous nodules. On top of these crags the welding fabric is contorted, indicating deformation at a late stage of its development either by movement of entrapped gas or by slip of the tuff on its emplacement slope. The tuff is bedded at its top (3) and is overlain by coarse volcaniclastic cross-bedded sandstones, which locally (4) include layers of coarse-ribbed brachiopods typical of a shallow-water marine environment.

The route follows the edge of the steep blocky scree to (5). Here the contact between the top of the Pitts Head Tuff and the overlying sandstone is exposed, dipping southwards into the hill. The sandstone includes a thin basalt flow.

Higher on the slope (6), across the scree composed of large blocks, the sequence underlying the Lower Rhyolitic Tuff Formation is exposed.

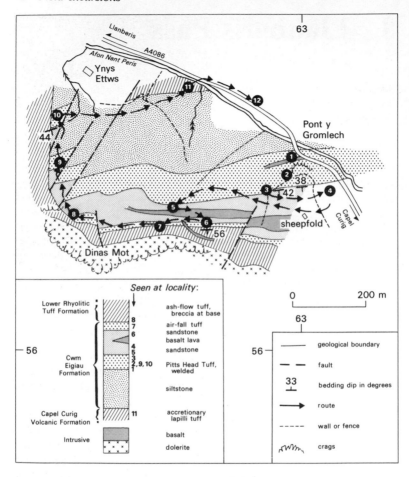

Figure 30   Sketch map. Excursion 8, Llanberis Pass.

Climb along the edge of the outcrop, where a weathered basalt flow is succeeded by cross and even-bedded sandstone and then by a prominent horizon of fine-grained air-fall tuff with locally a distinct sandy fraction. The tuff is conformably overlain by a coarse pyroclastic breccia, which occurs at the base of the Lower Rhyolitic Tuff Formation. This conformable sequence has been traced eastwards across the Llanberis Pass and occurs, around an anticlinal structure, in the slopes of the northern side of the pass. The pyroclastic breccia comprises blocks of vesiculated basalt, acid tuffs and sandstone in a matrix of shards and feldspar crystals.

Trace the base of the breccia westwards to (**7**). Here the underlying air-

Figure 31 Annotated photograph showing route, Pont y Gromlech (Excursion 8) (IGS photograph).

fall dust tuff has been quarried for honestone. In the cliffs above, the pyroclastic breccia, which shows marked variation in thickness, grades upwards into medium-grained acid tuff composed of closely packed shards. The tuff is intruded by a thick dolerite sill.

The conformable sequence continues westwards to (**8**). Here the breccia sharply downcuts the underlying strata; the junction is well exposed and is clearly not faulted. Further downslope, in the block-strewn ground, isolated outcrops (**9, 10**) indicate that the distinctive welded Pitts Head Tuff is downfaulted to the west, and the close disposition of the pyroclastic breccia indicates that these faults developed prior to its emplacement. The breccia transgressed the fault scarps down to the level, at least, of the Pitts Head Tuff.

The interpretation of this passage from a conformable to a transgressive base of the Lower Rhyolitic Tuff Formation is important with regard to the sequence in southern Snowdonia. Here, prior to the eruption of the Lower Rhyolitic Tuff Formation, a thick wedge of slump breccias occurred. These have been interpreted as developing when upward movement of magma in the underground reservoir disturbed the accumulated material in the depositional environment. This movement is probably reflected in the Llanberis Pass in the development of the sub-breccia fault scarps.

However, here the breccia is clearly pyroclastic; it grades upwards into the normal tuffs of the formation and is an integral part of the formation, as in Cwm Idwal (Excursion 5).

Take the track back to the main road, and proceed to (11). Here, volcaniclastic sandstones and sandy tuffites, well below the Pitts Head Tuff, include fine samples of accretionary lapilli. The conspicuous small round lapilli (Fig. 3a) are believed to have formed by the accumulation of volcanic dust about falling raindrops. These beds are probably the representatives of the Capel Curig Volcanic Formation.

At the roadside, near Pont y Gromlech (12), are the Gromlech Boulders, much loved by the climbing fraternity. These consist of the basal pyroclastic breccia of the Lower Rhyolitic Tuff Formation, fallen from the crags above, on the northern side of the pass. Here the breccia can be examined closely in comfort.

9 Northeast of Capel Curig

Duration: ½ day.
Geological map: IGS 1:25 000 Capel Curig and Betws y Coed (SH 75).

This excursion is to examine the ash-flow tuffs of the Lower Crafnant Volcanic Formation northeast of Capel Curig village. These tuffs form the ridge extending southwards from the Llyn Cowlyd area, which is particularly prominent when viewed from the west between Llyn Ogwen and Capel Curig. To reach the ridge from the Capel Curig road junction a convenient footpath crosses the underlying strata, which are locally well exposed and include a number of interesting volcanic and fossil localities. The route is easy, and at the upper part of its course is one of the vantage points for the classic view of Snowdon: across Llynnau Mymbyr to the west.

The route (Figs 32, 33) begins at Capel Curig village, around which limited parking is available.

From the road junction [721 581] at Capel Curig, follow the marked public footpath at the side of the church along the rising meadow to Curig Hill. Close to an old wall (1) a massive silicified acid tuff, dipping at 70–80° to the east, forms a prominent feature. This tuff is overlain by bedded coarse-grained basic tuff composed of basaltic and chloritised vesiculated glass fragments. These basic tuffs weather distinctively into the jagged crags at the hilltop referred to as the Pinnacles (Fig. 33). On Curig Hill the basic tuffs are at their thickest development, and to the north and south they thin out into a sequence of sandstones and thin acid tuffites, clearly featured to the north. At the Pinnacles the tuffs dip into a central point and are interpreted as infilling the upper part of a small volcanic vent that developed at this locality as the associated sandstones were being deposited. It is possible in the basic tuffs at the top of Curig Hill to distinguish slumped bedding, where the extruded tuffs collapsed towards the open neck of the vent.

Traverse northwards to the footpath and (2). Here, thin beds of fine-grained acid tuffite break the sandstone sequence. The sandstones are medium to coarse grained, cross-bedded and fossiliferous, with, in particular, coarse-ribbed dinorthid brachiopods. The fine-grained tuffites

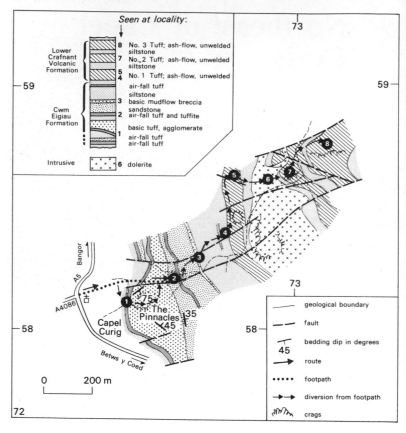

Figure 32 Sketch map. Excursion 9, northeast of Capel Curig.

reflect dust-rich and probably distant eruptions, the products of which settled into the sea. Such horizons are common in this part of the sequence.

Follow the footpath through the gate, then immediately turn left, uphill; pass through another gate on to the open hillside. Higher on the slope (**3**), at the top of the sandstones, a 3 m band of basic tuff is overlain by a mudflow breccia. The basic tuff is soft weathered, and the breccia developed through the slumping of previously deposited material, probably stimulated by earthquake shocks associated with the volcanic activity.

Cross the gorse-covered slope to the prominent feature formed by the No. 1 Tuff of the Lower Crafnant Volcanic Formation (**4**). Here the base of the tuff is exposed, overlying cleaved blue–grey siltstones, which include a thin fine-grained air-fall dust tuff. The basal part of the No. 1 Tuff is rich in feldspar crystals and lithic clasts and is well cleaved. Included fragments

Figure 33 Annotated photograph, the Pinnacles, Curig Hill (Excursion 9).

of brachiopods are evidence to show that the acidic ash-flow was emplaced in a submarine environment. The tuff is unwelded and shows a crude upward grading. Its central part (5) consists of shards, feldspar crystals and pumice clasts, and it passes up to fine-grained silicified vitric dust tuff at the top. Irregular massive bedding in the central part indicates that the flow at this point was beginning to lose its momentum.

The No. 1 Tuff is overlain by siltstone, which is intruded by dolerite (6). Here the dolerite is coarse grained, and the characteristic interlocking of feldspar and ferromagnesian crystals (**ophitic** texture) can be seen with the naked eye. The dolerite thickens into the large crag feature to the south and wedges out to the north.

The No. 2 Tuff (7) has a distinctly chloritic base, which passes up into a clean fine-grained acidic vitric tuff with few feldspar crystals. It is characteristically silicified, and patches of siliceous nodules occur at its top. The tuff retains these characters over most of its outcrop.

A detailed study has shown that the No. 1 and No. 2 Tuffs were erupted in central Snowdonia at the time of the accumulation of the uninterrupted pile of ash-flow tuffs comprising the Lower Rhyolitic Tuff Formation on Lliwedd (Excursion 14). The No. 1 and No. 2 Tuffs represent large eruptions that escaped eastwards into a marine environment. Their shard-rich composition with subordinate crystals and lithic clasts is typical of ash-flows. Their fairly constant thickness over an area of some 50 km^2, and ill-defined sorting combined with overall upward grading, indicate primary flow and rapid emplacement. The lack of induration of the included clasts, the preservation of fossil fragments and the absence of welding indicate that the flows lost most of their heat during transport. The bedding suggests water suspension, its presence indicating loss of energy within the main flows with the formation of subflows.

The No. 3 Tuff (**8**) is typically coarse grained and heterogeneous and, unlike the lower two tuffs, contains no feldspar crystals. Included large blocks of older tuff can be determined on weathered surfaces near its base, and a band of well-rounded blocks with little matrix occurs near its top. Shards are particularly coarse, up to 4·5 mm across, although distinctive cuspate margins are well developed. Unlike the lower two tuffs, the No. 3 Tuff is interpreted as being emplaced from a source to the north, and its characters suggest that the flow was less mobile.

This locality affords a fine view of the panorama to the east and southeast, across the valley of Afon Llugwy, extending from the Denbigh Moors, the Migneint and the Arenigs to the Penamnen Valley south of Dolwyddelan and the Moelwyns. To the southwest, Moel Siabod dominates the skyline. On retracing the path back to Capel Curig the classic view of the Snowdon massif across Llynnau Mymbyr dominates the western skyline.

10 Moel Siabod

Duration: ½ day.
Geological maps: IGS 1:25 000 Capel Curig and Betws y Coed (SH 75); and
IGS 1:25 000 Geological Special Sheet, Central Snowdonia.

This excursion is to examine an outlier of the Lower Rhyolitic Tuff
Formation on the southern flank of Moel Siabod (Fig. 34). The sequence
has been correlated with part of the Lower Crafnant Volcanic Formation at
Capel Curig (Excursion 9) and with part of the Lower Rhyolitic Tuff
Formation at Dolwyddelan (Excursion 11). The sequence in the outlier
comprises a Lower and an Upper Acid Tuff, separated by siltstones and
sandstones with a persistent basic tuff. The sequence is folded into an open
asymmetric syncline, with an axial plane trace trending between eastwards
and northeastwards across the outcrop.

The route is generally easy and provides panoramic views across the
wooded Lledr Valley to the east, and of the serrated profile across Yr

Figure 34 Moel Siabod viewed from south of the Lledr Valley (Excursion 10). The
walls of the high cwm are composed mainly of dolerite (IGS photograph).

Arddu and the Moelwyns to the southwest. The area is reached by the Forestry Commission road leading northwards from the A470 [744 528]. This road, clearly marked on the 1:50 000 sheet, is gated, and permission to drive along it must be obtained from the commission office at Gwydyr Uchaf, Llanrwst.

Take the main track northwards, turning left beyond the bridge crossing Afon Ystumiau, to the car-parking area, where the route commences (Fig. 35).

From the car park, follow the track crossing the tributary of Afon Ystumiau on to the footpath leading past a small quarry. Just beyond the quarry, leave the footpath and bear southwestwards to the base of the crags (1). Here, on the southern limb of the syncline, the base of the Upper Acid Tuff downcuts through the underlying sediments into basic tuff. The acid tuff is rich in feldspar crystals, and upwards on the crags a faint bedding lamination can be determined.

Walk westwards to the stream and then upslope to (2) on the northern limb of the syncline. Here the base of the Upper Acid Tuff is not exposed, although its limit is defined by a pronounced feature extending to the southwest. In these crags an upward transition can be recognised, from ill-defined massive beds with faint internal lamination, to well-defined massive beds (up to 2 m) and flaggy beds (up to 0·5 m). With this transition there is an increase in the epiclastic fraction in the tuffs, and sedimentary structures such as cross-bedding, channelling and the local deformation of bases can be seen. This tuff is unwelded and has been correlated with the No. 1 Tuff of the Lower Crafnant Volcanic Formation at Capel Curig. Its source lay in the Snowdon area, where the Lower Rhyolitic Tuff Formation is represented by a thick unbedded sequence of ash-flow tuffs (Excursion 14). Here, at Moel Siabod, the sequence comprises a composite unit of small primary ash-flows and pyroclastic flows developed from the slumping of previously emplaced deposits nearer the source area.

To the west (3) the Lower Acid Tuff crops out in the steep slope, which can be traversed fairly easily. It conformably overlies siltstones and fine sandstones and is well bedded. Through the cliffs a sequence can be determined from well-bedded tuff rich in feldspar crystals at the base (3), overlain by flaggy-bedded laminated silty tuffites showing graded bedding, which pass upwards with a gradual increase in the crystal content into crystal tuffs. The top of the laminated tuffite is locally deformed and disturbed by lobate intrusion of the overlying crystal tuff. In the crystal tuffs at the top of the sequence (4) rounded rhyolite clasts are randomly distributed. The sequence is interpreted as an accumulation of turbid flows and slumps from previously deposited pyroclastic debris.

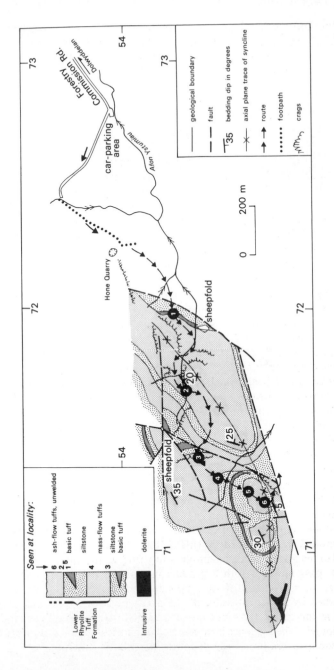

Figure 35 Sketch map. Excursion 10, Moel Siabod.

The overlying sedimentary strata are well featured in the grassy slopes on the northern limb of the syncline; and a bed of basic tuff correlated with that at (1) is exposed near the centre of the sequence. To the south (5) the Upper Acid Tuff is exposed, in a line of small crags, dipping southwards into the steep scarp (6), which lies close to the axial plane of the asymmetrical syncline. The traverse between these localities demonstrates an irregular, locally agglomeratic base overlain by well-bedded shardic tuff, which becomes progressively more muddy towards the top. The lower parts are generally massive, with crude upward grading and few cross-bedded silty laminations. The upper part comprises flaggy alternations of cross-bedded tuffites, with channel features, and thin, evenly bedded, air-fall, vitric tuffs.

Continue downslope across the lower tuff, and return eastwards to the Forestry Commission road and Dolwyddelan.

11   Dolwyddelan

Duration: ½ day.
Geological map: IGS 1:25 000 Capel Curig and Betws y Coed (SH 75).

This excursion is to examine the Snowdon Volcanic Group, which is exposed in a tight ENE-trending syncline at Dolwyddelan in the Lledr Valley. It is one of the few areas where there is a complete sequence through the group, although, when it is compared with the central Snowdonia sequence (Excursion 14), there is clearly a much reduced thickness and contrasting lithological types. The preservation of the sequence results largely from the structure. The trend of the syncline differs from the northeastward trend of central Snowdonia, and the reasons for this difference have been the cause of much discussion.

The excursion begins at Dolwyddelan Castle, where there is a convenient car park [724 524] (Fig. 36).

Take the footpath northwards from the car park past Bryn Tirion Farm, where permission should be obtained to visit Chwarel Ddu (4). Continue on the footpath, across the inverted northern limb of the syncline, to a small waterworks hut (1). In this area the Lower Rhyolitic Tuff Formation comprises three thick acid tuff units, each of which forms a prominent eastward-trending feature. The lowest unit, at the base of the Snowdon Volcanic Group, is well exposed in the most northerly feature about 100 m northwest of the hut. It consists of bedded clast-rich tuffs, with no apparent grading, and a central siltstone layer. It is restricted to Dolwyddelan and Moel Siabod (Excursion 10) and does not extend to the southern limb of the syncline. The tuffs of the unit are mass-flow deposits, probably from the slumping of previously emplaced pyroclastics. As the succession is overturned, the rocks, although dipping to the north, are younger to the south.

The middle and upper units can be examined in the two features to the south, beyond a grassy depression within the intervening siltstones. They are of bedded unwelded acid tuffs showing an upward grading, with prominent lithic clasts near the base of the beds and pumice clasts above. Large infilled scour structures are well exposed on the inverted base of the upper unit, and the top shows worm tube casts. These two units have been correlated with the lowest tuffs of the Lower Crafnant Volcanic Formation at Capel Curig (Excursion 9) and are considered to have been emplaced as

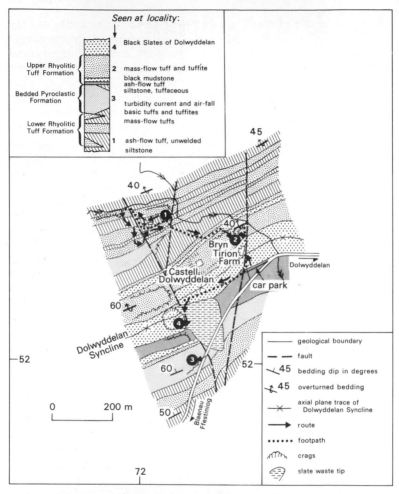

Figure 36   Sketch map. Excursion 11, Dolwyddelan.

primary ash-flows from a source in central Snowdonia (Excursion 14). The bedding here at Dolwyddelan suggests water suspension near the distal part of the flow, probably resulting from a loss of energy within the main flow, giving rise to the formation of subflows.

The features formed by these tuffs clearly indicate differential erosion by ice. To the north, in the flat peaty depression, the underlying sediments have been deeply scoured and smoothed off, whereas the tuffs form upstanding ice-scoured crags along strike. From these features the massive rocky profile and southern face of Moel Siabod to the north, and the distinctive ridge about Yr Arddu to the west, are particularly prominent.

Retrace the path back towards the car park, across the depressional feature formed by the scouring of the soft weathered Bedded Pyroclastic Formation, which on the limb of the syncline is poorly exposed. In the crags (2) between Bryn Tirion Farm and Dolwyddelan Castle, the Upper Rhyolitic Tuff Formation is well exposed. The formation comprises a heterogeneous admixture of acid tuff and mudstone. The pyroclastic content predominates, with lenses and more regular bands of tuff within tuffaceous black mudstone. The constituents are similar to those occurring in the Upper Crafnant Volcanic Formation and suggest that the formation was emplaced by the extensive downslope slumping of acid tuff, which incorporated, and was finally redeposited upon, unlithified mud.

Continue along the track back to the car park. From here it is possible for a small party to follow the main road to examine the Bedded Pyroclastic Formation on the southern limb of the syncline in the steep wooded slope (3) above the road. The formation consists of well-bedded basic tuffs, tuffites and sandy siltstones. Many beds show good normal or reverse grading, whereas others are cross-bedded. The tuffs represent both flow and fall deposits, with reverse grading in pumiceous tuff caused by the retarded settling through water of the light pumice fragments. Deposition in a marine environment is indicated by the current bedding and associated fossiliferous siltstones.

Return to the car park and take the path across the waste tips to the Chwarel Ddu Quarry (4). The quarry is excavated in the Black Slates of Dolwyddelan, overlying the Upper Rhyolitic Tuff Formation in the core of the syncline. The slates are black and graptolitic, with pyrite along the bedding and cleavage. Their junction with the underlying tuff, exposed on the northern side of the quarry, is disturbed by thrusting.

The limbs of the fold in this vicinity are parallel, the structure thus being termed an **isocline**, and they dip moderately northwards. All the rocks of the structure have been strongly deformed, and on cleavage surfaces a lineation is produced by the parallel preferred orientation of minerals and the long axes of pyritous spots and clasts. The attitude of cleavage in the limbs indicates that the fold was not originally isoclinal and that a subsequent phase of deformation tightened the structure.

Return to the car park at Dolwyddelan Castle.

12 Sarnau, North of Betws y Coed

Duration: ½ day.
Geological map: IGS 1:25 000 Capel Curig and Betws y Coed (SH 75).

This excursion is to examine the lithologies and distinctive sedimentary structures in the Middle and Upper Crafnant Volcanic Formations near Betws y Coed. The route lies entirely within the Llanrwst Mining District, where lead and zinc ores were extracted mainly during the years between 1848 and 1914. Evidence of this activity, in headings of abandoned mines, dumps, trial pits and openworks, occurs throughout the area. The excursion is within Forestry Commission ground, and access is almost entirely by foot along commission tracks. The routes are easy, and in addition to the pleasant silence of the forests there are many fine panoramic views.

The excursion route (Fig. 37) begins at the junction of the Forestry Commision track with the main road at Sarnau [775 590], where there is adequate parking space.

Southwest of the junction, massive blocky tuff of the Middle Crafnant Volcanic Formation crops out in the roadside at (1). Blocks of tuff and tuffite are included in a fine-grained silicified vitric tuff, which on its weathered surface is distinctively speckled by lighter 'spots'. These represent small patches of carbonate, which segregated during recrystallisation. Some of the tuffite blocks show remarkably irregular margins, which indicate that they were incorporated into the tuff while only partially lithified. From this locality the well-developed dip and scarp features in the formation are clearly displayed east of the road. These are produced by the differential erosion of resistant silicified tuffs and tuffites and of softer cleaved blue–black pyritous mudstones and siltstones. To the south the line of old mine dumps marks the site of the old lode working.

On the east of the road, skirt the tuff scarp feature to (2), noting the well-defined negative feature developed in the underlying mudstones to the south. In places along the scarp the beds overlying the blocky tuff are well exposed. These consist of banded, fine-grained, light grey tuff and dark blue–grey mudstones, which represent the settling of fine volcanic dust and sediment through the water after the emplacement of the blocky ash-flow. Locally, these striped horizons are contorted into tight folds, and the

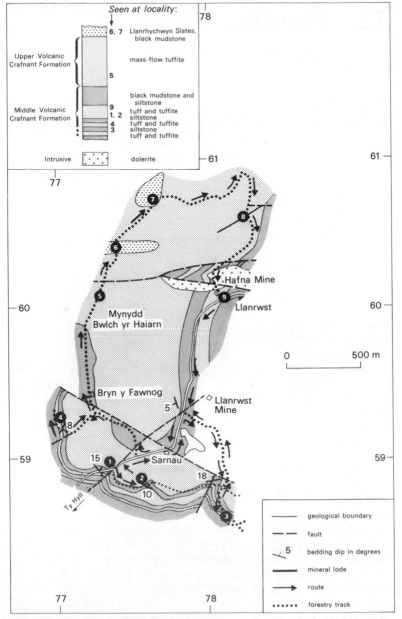

Figure 37 Sketch map. Excursion 12, Sarnau.

overall appearance is suggestive of flow-banded rhyolite (Fig. 38). However, these rocks are clearly not magmatic. They are interpreted as beds that have reacted to shock, either by earthquakes, which would have been common in a volcanically active area, or by the rapid emplacement of an overlying ash-flow tuff, while the beds were still unlithified. The shock caused migration of the included water, and the even banding was deformed into these internal folds; in some instances the banding was totally obliterated, effecting a complete admixture of pyroclastic and epiclastic materials. Beds with such convolute laminations are common in the Middle Crafnant Volcanic Formation in this area.

Return to the road and follow it northeastwards. Take the forestry track from the Sarnau Outdoor Pursuits Centre [778 593], near the derelict heading of Llanrwst Mine, and follow it southeastwards to the large trackside cutting (3). This exposure is typical of the Middle Crafnant

Figure 38 Convolute laminated tuff, tuffite and mudstone (2, Excursion 12). Height of specmen is 30 cm (IGS photograph).

Volcanic Formation in this area, comprising evenly bedded tuff, tuffite and mudstone. In this section it can be seen that the background sedimentation of the formation was of deep-water blue—black muds but that this was affected by the products of distant volcanic activity. These encroached upon the sedimentary environment by distinct ash-flows, by the slumping of previously emplaced pyroclastic debris that incorporated much sediment during transport, and by air-fall and water-settled fine vitric dust and feldspar crystals. No shallow-water sedimentary structures are seen, and close examination, near the base of the section, will provide evidence of thin sandstone bands deposited from turbidity currents.

Return to the main road at [775 590]. Follow the forestry tracks northwards towards Bryn y Fawnog, then westwards to the trackside cutting (4). Here again the even-bedded character of the formation is clearly apparent. A blocky ash-flow tuff at the top of the section overlies a flaggy sequence of tuffite, siltstone and mudstone. Within some of the beds underlying the blocky tuff, evidence of post-depositional yielding is well displayed, with 'flames' of banded mudstone extending upwards into the overlying beds. This locality presents a fine panorama of Snowdonia to the west across the ridge above Llyn Geirionydd, including part of the Snowdon ridge, the Glyders and, most prominently, the Carneddau.

Return to the main track, near Bryn y Fawnog, and follow it northwards on to the scarp of the Upper Crafnant Volcanic Formation. In a small roadside exposure (5) the tuffite is characteristically unbedded and dark blue, with scattered feldspar crystals and coarse shards. The crystals and shards are typical of acidic ash-flow eruption; however, the matrix consists of fine greenish mica flakes and chlorite, more typical of mudstone. The admixture of components is interpreted as having developed from the incorporation of unlithified mud into a large ash-flow. In this area the tuffite is widespread, creating the afforested plateau east and northwest of Sarnau.

To the north the track continues on to the dip slope feature at the top of the formation, where the pyroclastic element is less and the rock is cleaved blue tuffaceous slate. Along the roadside the overlying Llanrhychwyn Slates, blue—black with small pyritic spots, are exposed in two small outliers (6, 7) on the dip slope feature. Just north of (6) the underlying tuffaceous slates include large rounded pumice blocks.

Northeast of Mynydd Bwlch yr Haiarn the track bears southwards along the top of the steep scarp and crosses the open lode workings (8) in the Upper Crafnant Volcanic Formation in the Hafna Mine property. The lodes are along steeply inclined fault planes, and examination of the old waste tips will produce specimens of the ore minerals (galena and sphalerite) as well as the gangue minerals (quartz and calcite).

The track extends down to the roadside, below the Hafna Mine heading, where it transgresses into bedded acid tuff and tuffites of the Middle Crafnant Volcanic Formation, exposed below the dumps of tailings from the derelict mill (9).

Return by the narrow road, southwards to the Sarnau area.

13 Conwy Valley

Duration: 1 day.

This excursion is to examine the basic rocks of the Dolgarrog Volcanic Formation and Tal y Fan Volcanic Formation, and the closely associated acidic rocks of the Crafnant Vocanic Group, on the western side of the northern Conwy Valley. The variations in the basic volcanic rocks that are to be seen reflect their localised accumulation, and this contrasts sharply with the extensive contemporaneous acidic ash-flows of the district. The excursion is in three parts, with visits to localities in the Dolgarrog, Afon Porth-llwyd and Tal y Fan areas. As some access roads are narrow, it is most suitable for those using private vehicles.

(a) Dolgarrog

This part of the excursion is in Coed Dolgarrog at the foot of the steep scarp immediately west of Dolgarrog village. The valley is the site of the major Conwy Valley Fault, which separates the Silurian to the east from the Ordovician on the west. Southwards from Dolgarrog the valley profile shows a marked contrast, produced by glacial erosion, between steep crags in the resistant Ordovician volcanic rocks and gentler slopes in softer Silurian sediments. The valley was glacially overdeepened, and recent evidence indicates that the infilling glacial and alluvial deposits are up to 150 m thick east of Dolgarrog.

The route (Fig. 39a) begins at the small access road on the western side of the main road (B5106), 200 m north of Dolgarrog Post Office [7693 6742].

Take the footpath northwards towards Afon Porth-llwyd, and continue beyond the stile and rhyolite crags to the grassy bench [7669 6757] (1). The bench is the top surface of a white-weathering rhyolite, which, with a conformable base and top, is a sill intrusive into the Middle and Upper Crafnant Volcanic Formations. In the crags below, fine flow banding (Fig. 40) and incipient nodules can be examined.

Towards the scarp above the rhyolite there are two more prominent bench features, which can be traversed due west of (1). The lower comprises interbedded thin beds and laminae of dark grey siltstone and paler acid tuffite. Close inspection reveals delicate load cast and disturbed bedding structures. The feature above is formed by an acid ash-flow tuff about 5 m thick, which superficially resembles the rhyolite below. However, it is an unwelded shardic tuff with numerous clasts of indurated siltstone and silty tuffite, patchy concentrations of feldspar crystals, and pumice scattered throughout. Its heterogeneous character is accentuated by differential silicification.

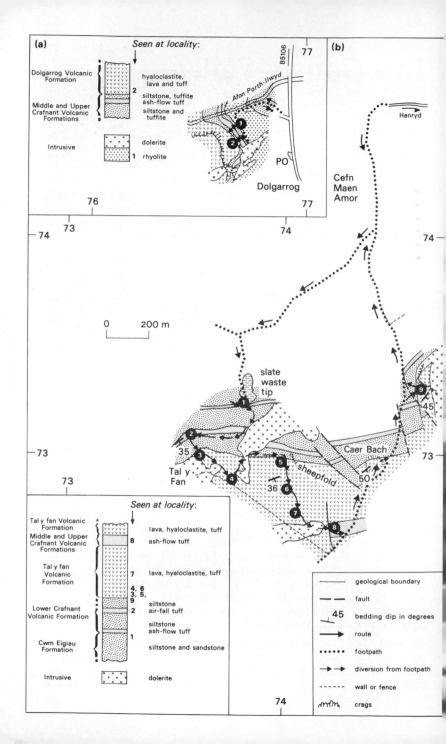

(a)

Seen at locality:

Dolgarrog Volcanic Formation		hyaloclastite, lava and tuff
		siltstone, tuffite ash-flow tuff
Middle and Upper Crafnant Volcanic Formations		siltstone and tuffite

dolerite

1 rhyolite

(b)

Henryd

Cefn Maen Amor

Dolgarrog

Afon Porth-llwyd

B5106

PO

slate waste tip

Tal y Fan

Caer Bach

sheepfold

77

76

77

74

73

73

74

73

74

0   200 m

Seen at locality:

Tal y fan Volcanic Formation		lava, hyaloclastite, tuff
Middle and Upper Crafnant Volcanic Formations	8	ash-flow tuff
Tal y fan Volcanic Formation	7	lava, hyaloclastite, tuff
	4, 6, 3, 5,	
Lower Crafnant Volcanic Formation	9	siltstone air-fall tuff
	2	siltstone ash-flow tuff
Cwm Eigiau Formation	1	siltstone and sandstone

Intrusive

dolerite

——	geological boundary
– –	fault
⤙45	bedding dip in degrees
⟶	route
••••	footpath
→→	diversion from footpath
– – –	wall or fence
⌢⌢⌢	crags

Figure 40 Contorted flow banding in rhyolite (Excursion 13a) (IGS photograph).

The overlying basic volcanic rocks of the Dolgarrog Volcanic Formation crop out at the base of the scarp above, where small blocks and fragments of basalt and feldspar crystals are dispersed in a matrix of fragmental chloritised glassy vesiculated basaltic lava.

From (1) continue southwards some 100 m along the path before bearing west by the large fallen blocks, to the crags (2) in the basal facies of the formation. There the base contact with silty acid tuffite is seen at the northern end of the crag, although along the main face the contact is intruded by dolerite. The basal rock is a breccia comprising basaltic lava blocks, up to 0·5 m, and fine-grained pillows with chilled margins set in a crystal rich hyaloclastite matrix. The contrasting textures in the blocks and pillows, and their juxtaposition, suggest that the rocks may not have formed *in situ* but may have been a secondary flow from a nearby source area.

(b) Pont Newydd on Afon Porth-llwyd

North of Dolgarrog at Tal y Bont [7668 6885], take the road westwards up the valley scarp towards Llyn Eigiau. At the junction, some 2 km along,

Figure 39 Sketch maps, Excursion 13, Conwy Valley: (a) Dolgarrog (Excursion 13a); (b) Tal y Fan (Excursion 13c).

turn left on to the road down to Afon Porth-llwyd. A parking area is available just beyond Pont Newydd.

In the stream bed, particularly downstream from the bridge, there are some excellent washed surfaces of pillow hyaloclastite, in the Dolgarrog Volcanic Formation. Fragmented basaltic pillows, in which vesicles are concentrically arranged parallel to original pillow surfaces, are supported in a matrix of dark green chloritised comminuted vesiculated basic glass (Fig. 41). Crags by the stream are in pillow and massive hyaloclastite.

(c) Tal y Fan

The excursion begins at the mountain gate [7448 7461] (Fig. 39b) 2·5 km due west of Henryd, where cars should be carefully parked, to allow vehicular access to the mountain track.

Follow the track southwestwards past the standing stone before turning southwards up to the quarry on the flank of Tal y Fan. Towards the quarry the crags are formed of deltaic sandstones in the Cwm Eigiau Formation. The quarry (1) is in grey siltstones with interbedded fine sandstones, which dip moderately southwards. The rocks are well cleaved and apparently provided rough roofing slates.

The siltstones are overlain by an acid tuff, which locally varies in

Figure 41 Pillow breccia. Blocks of broken basaltic pillows and lava toes in a hyaloclastite matrix, Pont Newydd (Excursion 13b) (IGS photograph).

thickness between 5 and 10 m. Just southeast of the quarry the upper part of the tuff is exposed in minor outcrops, where it is fine-grained and thinly bedded. Above the quarry the tuff is faulted out, but the more massive basal part of the tuff is exposed along strike to the west. This is the lowermost ash-flow tuff of the Lower Crafnant Volcanic Formation – part of the No. 1 unit, which is some 30 m thick to the southwest in the Capel Curig area (Excursion 9).

Proceed southwards from the quarry to the crags running northeast– southwest, which are at the margin of a transgressive dolerite body – an offshoot of a complex intrusion into the local sequence. At 150 m south of the quarry turn westwards along the rise to the exposures just north of the peat depression (2). These are of fine air-fall dust tuffs interbedded with siltstone tuffite – the lateral equivalent of the No. 2 unit of the Lower Crafnant Volcanic Formation (Excursion 9).

There are fine views of the North Wales coast from this location. To the northwest the Penmaenmawr microgranodiorite intrusion forms the high ground about the conspicuous upstanding pillar remnant in the hilltop quarry. The Conwy Rhyolite Formation forms the coastal feature extending down to Conwy. Eastwards to the Conwy estuary, there is a sharp contrast between the open grazing land of the Ordovician and the fenced lower land in the Silurian.

Just south of (2), columnar-jointed dolerite crops out at the edge of the peat flat, and the feature extending eastwards marks the base of the Tal y Fan Volcanic Formation (3). The rock is massive hyaloclastite, greyish green in colour, with conspicuous scattered clasts and feldspar crystals. Close examination shows its clastic texture, composed of small fragments of green chloritised vesiculated glass with paler secondary minerals, feldspar, quartz and calcite, locally abundant.

The next locality (4) is some 200 m to the southwest, about the minor bend in the stone wall on the hill crest. The hyaloclastite contains isolated large blocks of vesiculated lava and pillows, which in places are concentrated, forming a breccia facies. The dolerite (above 1) extends into the basic volcanic rocks here, and it is an interesting exercise to trace the boundary between the magmatic and clastic rocks of similar overall composition.

Above a derelict sheepfold approximately 250 m to the ENE, the base of the formation is marked by a strong feature. The adjacent depression is occupied by underlying siltstones. Joint faces here (5) provide a section through the massive hyaloclastite, which is paler than to the west, and bedding is discernible. Bedding becomes more apparent upwards, and 60 m to the south mixed epiclastic and volcaniclastic material forms well-defined beds. In this vicinity the sequence includes lenses of very fine-grained acid air-fall tuff.

Proceed downslope to the minor knolls (6) in the east–west depression. They comprise massive vesiculated basaltic lava; and in adjacent small outcrops, blocks and pillows of this lava occur within a dark green hyaloclastite matrix.

The route continues southwards through a break in the east–west feature, which marks an overlying massive hyaloclastite. Faint bedding indicates that it is steeply inclined. The base of overlying breccias (7) transgresses this bedding. In the breccia, clasts up to 1 m across are largely of basic volcanics, but a few indurated sediment and acid tuff clasts suggest that the breccia is a block-flow from a nearby source that has downcut the hyloclastite.

Exposures (8) 100 m north of the junction of track and mountain wall are in a distinctive unwelded acid tuff, some 50 m thick, of the undivided Middle and Upper Crafnant Volcanic Formations, which extends through the middle of the Tal y Fan Volcanic Formation. The lower part is rich in clasts, amongst which indurated black mudstone is prominent, and it fines upwards to a porcellaneous top. Overlying this acid tuff, the upper part of the Tal y Fan Volcanic Formation – comprising discontinuous beds of lava, hyaloclastite, basic tuffs and black mudstone – extends down the hill towards Roewen. From this area there are fine views southwards, towards the Crafnant country and up the Conwy Valley.

Continue northeastwards along the track, pausing at the Caer Bach hill fort, situated on the lowest rocks of the Tal y Fan Volcanic Formation. Blocks of mudstone are incorporated into the hyaloclastites at the base; and in the minor outcrops below the fort, bedding is less massive than to the west.

Follow the track northwards for 350 m before bearing eastwards up slope to the crags (9) near the margin of the dolerite forming the higher ground. The crags are of grey–green, flaggy and thinly bedded, basic tuffs locally showing graded bedding. They comprise finely comminuted hyaloclastite fragments and are the lateral equivalents of the massive breccia hyaloclastite seen at the base of the formation to the west. The latter are near-source accumulations, whereas the thinly bedded tuffs represent peripheral deposits resulting from collapse of the unstable central pile and deposition from turbidity currents.

The path continues northwards across a minor stream to rejoin the track back to the starting point.

14 Snowdon: Pyg and Miners' Tracks

Duration: 1 day.
Geological map: IGS 1:25 000 Geological Special Sheet, Central Snowdonia.

Both these routes start at the Pen y Pass Youth Hostel at the top of the Llanberis Pass. They are probably the most popular routes to the Snowdon summit, and certainly the most impressive. To most people who set out on these tracks the aim is, understandably, to reach the top of the mountain. As a result the presentation of the excursion differs from the others given in this book in that it refers to the general geological aspects of the walk rather than to detail at specific localities. Both routes are marked on the Ordnance Survey 1:50 000 Sheet 115.

(a) Pyg Track

From Pen y Pass, the Pyg Track skirts the northern side of Tal y Llyn ridge to the col at the eastern end of Crib Goch. Near Pen y Pass the track crosses well-cleaved acid tuffs of the Lower Rhyolitic Tuff Formation. These tuffs are remarkably homogeneous, shard rich and unwelded, containing few small chloritised pumice clasts; they are elongated parallel to the cleavage, and contain fewer fragmented feldspar crystals. Most distinctively, there is no clear indication of bedding in the tuffs. The overlying well-bedded basic tuffs and tuffites of the Bedded Pyroclastic Formation are best exposed high on the Tal y Llyn ridge, although downfaulted exposures occur in places along the track. As the track rises to the col it crosses flow-banded rhyolite intrusions in the massive acidic ash-flow tuffs. The rhyolite is fine-grained and dark blue−grey, although white weathered with ochreous-stained joint surfaces. Flow banding is locally highly contorted, and zones of brecciation, columnar joints and siliceous nodules are common.

At the col, where the Crib Goch track joins the Pyg Track, the overlying basic tuffs and tuffites of the Bedded Pyroclastic Formation are well exposed. Here, near the base of the sequence, fine acidic air-fall dust tuffs and few massive blocky ash-flow tuffs occur. This position affords a fine

Figure 42   Northern ridge of the Snowdon Horseshoe viewed from south of Llyn Llydaw (Excursion 14a). From left to right: Snowdon, Crib y Ddysgl and Crib Goch. The Miners' Track, from the edge of Llyn Llydaw up to Glaslyn, is clearly visible (IGS photograph).

view across Llyn Llydaw to the impressive crags of Lliwedd on the southern ridge of the Snowdon Horseshoe. These crags are composed of unbedded cleaved ash-flow tuffs of the Lower Rhyolitic Tuff Formation, in marked contrast to the equivalent bedded sequences at Cwm Idwal (Excursion 5) and Capel Curig (Excursion 9). The contact with the overlying bedded basic tuffs is clearly discernible on the southeastern flank of the peak.

Westwards into the Snowdon Horseshoe, the track crosses bedded basic tuffs and tuffites near the base of the Bedded Pyroclastic Formation. High on the northern side of the track the jagged outline of the Crib Goch ridge (Fig. 42), flanked by steep blocky screes, dominates the skyline. The ridge lies within a large rhyolite intrusion, and even at a distance its contact with the soft weathered bedded basic tuffs can be distinguished.

In places the track crosses into the underlying massive cleaved acid tuffs, locally bedded at the top, and north of Glaslyn a steeply inclined rhyolite intrusion extends from the ridge west of Bwlch Goch to the south of the path. From this locality [618 549] the upper ice-scoured rocky cwm of Glaslyn, with the precipitous crags below the Snowdon summit in its back wall, is well displayed. In these crags it is possible to determine the contact between the Lower Rhyolitic Tuff and Bedded Pyroclastic Formations and the Snowdon Syncline.

Further west the track continues close to the junction of these two formations to the old open workings of the Britannia Mine, which was mainly developed in the middle of the nineteenth century. The copper mineralisation is closely associated with the boundary between the acid and basic tuffs. Above the mine the track climbs through the well-bedded sequence of the Bedded Pyroclastic Formation, comprising basic tuffs, tuffites, thin basaltic lavas (locally highly vesicular) and tuffaceous sediments. In places the sediments are richly fossiliferous, with trilobite and coarse-ribbed brachiopod fragments being particularly distinctive.

At the ridge west of Crib y Ddysgl, the track joins the railway line, which it follows to the summit over exposures of the Bedded Pyroclastic Formation. With a certain element of luck for a clear day, the peak provides an extensive panorama, across Anglesey and Lleyn to the west, the Cambrian sequence of the Harlech Dome and the Ordovician volcanics of Cader Idris to the south, the Silurian of the Denbigh Moors to the east, as well as the full range of adjacent peaks in Snowdonia.

(b) Miners' Track

From Pen y Pass the Miners' Track skirts the southern side of Tal y Llyn ridge into the centre of the Snowdon Horseshoe. For most of its length, to where it joins the Pyg Track above Glaslyn, it crosses the massive cleaved ash-flow tuffs of the Lower Rhyolitic Tuff Formation. Its line extends along the lower parts of the three cwms contained within the Snowdon Horseshoe, with a wealth of localities showing the erosional and depositional effects of the ice. Just south of Pen y Pass the open end of the lowest cwm hangs high above Nant Gwynant. The track skirts the northern side of Llyn Teyrn, occupying the lowest cwm. It crosses the edge of a large well-exposed dolerite intrusion, with spectacular columnar joints in the crags on the southern side of the lake.

To the west the track climbs across the cleaved unbedded acid tuffs to the moraine-damned Llyn Llydaw in the middle cwm (Fig. 43). The grass-covered moraines are most extensive near the lip of the cwm, although small well-featured mounds extend on to the hillslope about the northeastern margin of the lake (Fig. 44). From this locality the Crib Goch ridge and its scree-covered south-facing slope, although foreshortened, are most spectacular. The track passes the derelict mill of the Britannia Mine on the northern shore of Llyn Llydaw. Above, ice-scoured crags in the massive tuffs of the Lower Rhyolitic Tuff Formation form the main part of the slope up to the Pyg Track. Thick quartz veins are locally distinctive, even at a distance.

Figure 43 View eastwards from near Snowdon summit across Glaslyn and Llyn Llydaw (Excursion 14b). Crib Goch ridge on the left and Moel Siabod in the upper centre. The Pyg and Miners' Tracks are clearly visible (IGS photograph).

Figure 44 Moraine, Llyn Llydaw (Excursion 14b) (IGS photograph).

To the west the track follows the valley of Afon Glaslyn, in the shadow of Lliwedd to the south, up into the highest cwm: Glaslyn. This rock-floored cwm, bound by steep ice-scoured crags and occupied by the spectacularly blue–green-coloured lake, is probably close to most people's idea of a glacial cwm. It drains across a pronounced rock lip into Llyn Llydaw below. On the northern side of the lake the well-defined path ends close to the waste tips from the Britannia Mine. The Pyg Track is reached by clambering up the rock slope, preferably a little east of the waste tips.

Appendix I   Glossary of geological terms

Terms used in the field excursion descriptions.

acid   Relating to igneous rocks containing more than 66% of silica.

accretionary lapilli   Small pea sized aggregations of volcanic ash that formed about nuclei, such as drops of moisture, during fall through the air following eruption from a volcano (Fig. 3a).

agglomerate   A volcanic rock formed of pyroclastic blocks or fragments of generally more than 50 mm diameter and which were ejected in a plastic state.

air-fall tuff   A tuff formed by the showerlike gravitational settling of pyroclastic debris from an eruption cloud (Fig. 2).

alluvial fan   A gently sloping apron of loose rock material, shaped in plan like an open fan, and deposited by a stream at a point where the gradient of the stream decreases abruptly. An alluvial cone is similar in form but slopes more steeply (Fig. 19).

ash   Fine fragmentary material, <2 mm in diameter, expelled during a volcanic eruption.

ash-flow   A turbulent mixture of pyroclastic debris and hot gas which flows in directions determined by the originating explosive volcanic eruption and by gravity.

ash-flow tuff   A tuff composed of pyroclastic debris deposited from an ash-flow.

axial plane   A plane that connects the hinges of all the individual folded beds within the fold. In a syncline it connects the troughs of each folded layer and in an anticline it connects the crests.

basalt   A fine-grained basic lava or minor intrusion composed mainly of calcic plagioclase and pyroxene, with or without olivine.

basic   Pertaining to igneous rocks containing less than 52% of silica.

bathyal   Pertaining to the marine environment between about 100 and 500 fathoms water depth.

bioturbation   The disturbance or churning of sediment by organisms.

boulder clay (till)   An unconsolidated and unstratified deposit of clay with boulders deposited beneath an ice sheet or glacier.

breccia   A coarse grained clastic rock composed of angular rock fragments set in finer grained matrix.

Caledonian orogeny   Major earth movements of Lower Palaeozoic age which reached their culmination at the end of the Silurian.

Caledonides   The folded mountain chain which resulted from the Caledonian orogeny; the Caledonian orogenic belt.

clast   An individual fragment of rock or mineral produced by the erosion or mechanical disintegration of a larger rock mass.

clastic rock   A rock composed of clasts.

cleavage Cleavage (in a rock) is a parallel planar fabric produced by deformation of the rock resulting in a tendency for the rock to split along closely spaced parallel planes (Fig. 7).

columnar joints Fractures which form by contraction during cooling of igneous rocks and disposed such that sets of the fractures form parallel prismatic columns within the rock. Most commonly found in lava flows and certain minor intrusions.

convolute lamination Contorted bedding laminae which are confined to a single well defined layer within a bedded sequence and are both overlain and underlain by parallel undisturbed layers (Figs 24 and 38).

cooling joints Fractures formed by contraction during cooling of igneous rocks.

cross-bedding An internal arrangement of the layers in a stratified sediment characterised by parallel sloping minor layers deposited at an angle to the principal stratification (Fig. 5).

desiccation cracks Cracks, usually in a polygonal system, formed on the surface of sand or silt when it is exposed to the air and dries out rapidly (Fig. 5).

de-watering The process whereby some of the water present between the particles of a newly deposited, subaqueous sediment is expelled in resopnse to increasing pressure caused by the continued accumulation of overlying sediment.

dolerite A medium-grained igneous rock generally forming minor intrusions and consisting mainly of calcic plagioclase and pyroxene, commonly with an ophitic texture.

epiclastic rocks Sedimentary rocks formed of fragments derived by the weathering and erosion of older rocks.

euhedral crystal A crystal displaying its natural faces without modification.

eutaxitic texture The texture formed in welded ash-flow tuffs when shards and pumice fragments are flattened and bent around crystals and lithic fragments during welding (Figs 2 and 3e).

extrusive igneous rocks Volcanic rocks formed by the eruption of material on to the surface of the Earth; includes lavas and pyroclastic rocks.

facies (sedimentary) The total lithological and palaeontological character of a rock from which its origin and environment of formation may be deduced.

fiamme Flattened pumice fragments in a welded tuff (Fig. 4).

flame structure Flame shaped intrusion of sediment, generally of mud, which has been squeezed upwards to protrude into the overlying coarser layer of sediment (Fig. 5).

flow bands Alternating layers of contrasting texture and/or composition in an igneous rock, formed as a result of flow in a magma (Figs 2 and 40).

flow front The front of an advancing lava flow.

flute mark A scour mark on a bedding surface formed by the action of a sediment-laden current of water flowing over newly deposited sediment. The mark has a deep up-current end from which it flares out and shallows in the down-current direction (Fig. 14).

fluvial sediment A sediment laid down by a stream or river.

fluvioglacial   Pertaining to the meltwater streams flowing from melting ice and to the deposits and landforms produced by the streams.

gangue   The uneconomic minerals in an orebody.

glacial striation   A groove or scratch on a rock surface caused by rock fragments embedded in the base of a moving glacier being dragged across the surface (Fig. 9).

glacial till   Unstratified glacial deposit, laid down directly from a glacier and not transported and redeposited by water from the melting glacier.

glass   See **volcanic glass**.

graded bedding   Internal structure of a clastic sediment whereby the maximum grain size progressively decreases from the base to the top of the bed (Fig. 5).

hyaloclastite   A deposit composed of comminuted basaltic glass formed by the fragmentation of the glassy skins of basaltic pillows or by the violent eruption of basalt under the sea.

intermediate   Relating to igneous rocks transitional between acid and basic.

intrusive igneous rock   An igneous rock formed from magma injected into the Earth's crust.

isocline   A fold, the opposing limbs of which are parallel.

joint   A fracture in a rock, usually planar and with little or no displacement across the fracture.

lapilli   Pyroclastic fragments in the range 2–64 mm ejected by volcanic eruption.

lithic   Pertaining to rock.

lithification   The process whereby unconsolidated material becomes converted to rock; the consolidation and induration of sediment or pyroclastic volcanic debris.

lithology   The general character of a rock; includes its composition, texture and primary and secondary structures.

littoral zone   The zone in a marine environment between low and high water marks; the intertidal zone.

load cast   A structure at the interface of two beds of sediment in which the sediment of the upper bed protrudes down into the sediment of the bed below, the upper sediment normally being the coarser (Fig. 5).

magmatic doming   Updoming of the Earth's surface caused by the generation and injection of magma below the surface.

mass flow   Downslope movement of sediment or rock material by gravity.

mudflow breccia   A rock consisting of angular clasts (many of mudstone), in a mudstone matrix and formed by the downslope collapse and mass flow of pre-existing, muddy sediment.

neritic zone   The marine environment between low tide mark and the outer edge of the continental shelf.

ophitic An igneous texture where prismatic plagioclase crystals are partially or completely included within pyroxene crystals.

periglacial Pertaining to processes and conditions at the periphery of glaciated areas where frost action is a significant factor.

perlitic texture Small-scale arcuate cracks caused by cooling in a volcanic glass (Fig. 2).

petrography The branch of geology dealing with the description and classification of rocks usually by means of microscopic examination.

pillow breccia A rock composed of fragments of broken pillow lava (Fig. 41).

pillow lava A rock mass composed of closely packed discontinuous spheroidal masses of lava (Fig. 2). Pillows most commonly form in basaltic lava extruded into a subaqueous environment.

plate tectonics A global tectonic model based on the proposal that the outer part of the Earth comprises a number of internally rigid but relatively thin plates, about 100 km to 150 km thick, which are continually in motion with respect to each other.

pumice A highly vesiculated rock composed of frothy glassy lava light enough to float on water; recrystallised fragments of pumice commonly occur in the tuffs of North Wales.

pyroclastic breccia A volcanic rock consisting of large angular fragments of material which have been fragmented, produced and erupted in a solid state by explosive volcanic activity.

pyroclastic rock A clastic volcanic rock composed of material fragmented and erupted by explosive volcanic activity.

quartzite A sandstone consisting almost entirely of quartz grains and normally cemented by interstitial quartz.

recrystallisation The formation of new crystalline mineral grains in a rock, the new mineral grains replacing some or all of the pre-existing mineral grains.

reduction spots Light coloured spheroidal patches within red and purple siltstones and mudstones produced by a local change in the content or oxidation state of iron within the spot (Fig. 8).

reverse grading Graded bedding in which the average size of the component grains increases from the base towards the top of a bed.

reworking In this account 'reworking' refers to the removal and redeposition by water of recently deposited tuff or sediment.

rhyolite An extrusive igneous rock of acid composition, commonly porphyritic and flow banded (Figs 3c and 40).

rock bar A low ridge of rock crossing a glacial valley.

shard A small fragment of glass having cuspate margins, frequently spindle shaped, and commonly found in pyroclastic rocks (Fig. 2).

slumped beds Beds distorted by subaqueous sliding or lateral movement of unconsolidated sediment.

stone polygons A polygonal arrangement of loose rock fragments in near-surface superficial material formed by freeze-thaw action (Fig. 10).

tuff   A lithified deposit of volcanic ash.

tuffite   A rock consisting of a mixture of pyroclastic (>25%) and epiclastic (>25%) fragments.

turbidite   A sediment deposited from a turbidity flow.

turbidity flow   A dense subaqueous flow of water and suspended sediment which moves downslope by gravity.

vent   An opening through which volcanic deposits are ejected or extruded.

vesicle   A small bubble-like cavity in a lava, formed by included gas.

vitric tuff   A tuff composed of volcanic glass fragments.

vitroclastic   Texture of a pyroclastic rock composed mainly of cuspate glass fragments (Fig. 3d).

volcanic glass   The amorphous non-crystalline product of the rapid cooling of a magma.

volcaniclastic   A clastic rock composed mainly of volcanic rock fragments.

volcanogenic sediment   A sediment whose origin is in some way associated with volcanic activity.

welding foliation   A parallel planar fabric formed by the flattening of hot plastic glass and pumice fragments during welding in a tuff (Fig. 4).

welded tuff   A tuff in which the hot, plastic pyroclastic fragments have been agglutinated under the influence of retained heat and the weight of overlying material (Fig. 4).

Appendix II References

Beavon, R. V., F. J. Fitch and N. Rast 1961. Nomenclature and diagnostic characters of ignimbrites with reference to Snowdonia. *Liverpool Manchester Geol. J.* **2**, 600–10.

Bromley, A. V. 1969. Acid plutonic igneous activity in the Ordovician of North Wales. In *The Precambrian and Lower Palaeozoic rocks of Wales*, A. Wood (ed.), 387–408. Cardiff: University of Wales Press.

Dakyns, J. R. and E. Greenly 1905. On the probable Peléan origin of the felsitic slates of Snowdonia, and their metamorphism. *Geol. Mag.* **5**, 541–9.

Harker, A. 1889. The Bala Volcanic Series of Caernarvonshire. Cambridge.

Howells, M. F., B. E. Leveridge, R. Addison, C. D. R. Evans and M. J. C. Nutt 1979. The Capel Curig Volcanic Formation, Snowdonia, North Wales: variations in ash-flow tuffs related to emplacement environment. In *The Caledonides of the British Isles – reviewed*, A. L. Harris, C. H. Holland and B. E. Leake (eds), 611–18. Spec. publ. Geol. Soc. Lond., no. 8.

Oliver, R. L. 1954. Welded tuffs in the Borrowdale Volcanic Series, English Lake District, with a note on similar rocks in Wales. *Geol. Mag.* **91**, 473–83.

Ramsay, A. C. 1881. The geology of North Wales. *Mem. Geol. Surv. G.B.* **3**.

Rast, N. 1969. The relationship between Ordovician structure and volcanicity in Wales. In *The Precambrian and Lower Palaeozoic rocks of Wales*. A. Wood (ed.), 305–35. Cardiff: University of Wales Press.

Rast, N., R. V. Beavon and F. J. Fitch 1958. Sub-aerial volcanicity in Snowdonia. *Nature* **181**, 508.

Williams, H. 1927. The geology of Snowdon (North Wales). *Q. J. Geol. Soc. Lond.* **83**, 346–431.

Wood, D. S. 1974. Current views on the development of slaty cleavage. *Rev. Earth Planet. Sci.* **2**, 369–401.

Appendix III Further reading

Volcanic rocks

Francis, P. 1976. *Volcanoes*. Harmondsworth: Penguin.

Institute of Geological Sciences, Geological Museum 1975. *Volcanoes*. London: HMSO.

Macdonald, G. A. 1972. *Volcanoes*. Englewood Cliffs, NJ: Prentice-Hall.

Sedimentary rocks

Greensmith, J. T. 1978. *Petrology of the sedimentary rocks*, 6th edn. London: Allen & Unwin.

Pettijohn, F. J. 1975. *Sedimentary rocks*, 3rd edn. New York: Harper & Row.

Tectonic structures

Hobbs, B. E., W. O. Means and P. F. Williams 1976. *An outline of structural geology*. New York: Wiley.

Glaciation

Embleton, C. and C. A. M. King 1975. *Glacial geomorphology*. London: Edward Arnold.

General

Challinor, J. 1978. *A dictionary of geology*, 5th edn. Cardiff: University of Wales Press.

Holmes, A. 1978. *Principles of physical geology*, 3rd edn. London: Nelson. This excellent book covers aspects of volcanic rocks, sedimentary rocks, tectonic structures and glaciation.

Whitten, D. G. A. and J. R. V. Brooks 1972. *The Penguin dictionary of geology*. Harmondsworth: Penguin.

Regional geology

Francis, E. H. and M. F. Howells 1973. Transgressive welded ash-flow tuffs among Ordovician sediments of N.E. Snowdonia. *J. Geol. Soc. Lond.* **129**, 621–41.

George, T. N. 1961. *North Wales*. British regional geology series, Institute of Geological Sciences.

Howells, M. F., B. E. Leveridge and C. D. R. Evans 1971. *Ordovician ash-flow tuffs in eastern Snowdonia*. Rep. Inst. Geol. Sci., no. 73/3.

Howells, M. F., E. H. Francis, B. E. Leveridge and C. D. R. Evans 1978. *Capel Curig and Betws y Coed: description of 1:25 000 sheet SH 75*. Classical areas of British geology series, Institute of Geological Sciences. London: HMSO.

Howells, M. F., B. E. Leveridge, C. D. R. Evans and M. J. C. Nutt 1980. *Dolgarrog: description of 1:25 000 sheet SH 76*. Classical areas of British geology series, Institute of Geological Sciences. London: HMSO.

Williams, D. 1930. The geology of the country between Nant Peris and Nant Ffrancon (Snowdonia). *Q. J. Geol. Soc. Lond.* **86**, 191–233.

Wood, D. S. 1969. The base and correlation of the Cambrian rocks of North Wales. In *The Precambrian and Lower Palaeozoic rocks of Wales*, A. Wood (ed.), 47–66. Cardiff: University of Wales Press.

Wood, D. S. 1974. Ophiolites, mélanges, blueschists and ignimbrites: early Caledonian subduction in Wales? In *Modern and ancient geosynclinal sedimentation*, R. H. Dott and R. H. Shaver (eds), 334–44. Soc. Econ. Palaeontol. Mineral. spec. publ., no. 19.

Index